Predicting Tillage Effects on Soil Physical Properties and Processes

ASA Special Publication Number 44

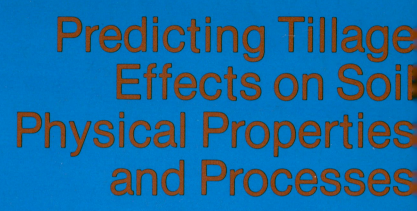

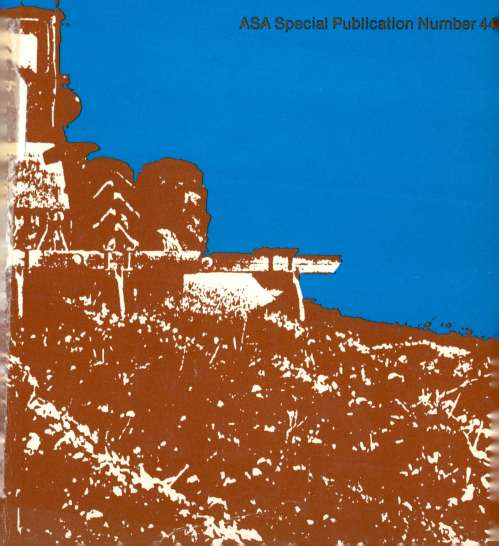

Predicting Tillage Effects on Soil Physical Properties and Processes

ASA Special Publication Number 44

Proceedings of a symposium sponsored by Divisions A-3,
S-6, and S-1 of the American Society of Agronomy
and the Soil Science Society of America.
The papers were presented during the annual meetings
in Detroit, Michigan, November 30–Dec. 5, 1980.

Organizing Committee
D. M. Van Doren—Chairman
R. R. Allmaras
D. R. Linden
F. D. Whisler

Editorial Committee
P. W. Unger—Co-editor
D. M. Van Doren, Jr.—Co-editor
F. D. Whisler
E. L. Skidmore

Managing Editor
David M. Kral

Assistant Editor
Sherri Hawkins

1982
AMERICAN SOCIETY OF AGRONOMY
SOIL SCIENCE SOCIETY OF AMERICA
677 South Segoe Road
Madison, Wisconsin 53711

American Society of Agronomy
Soil Science Society of America
677 South Segoe Road, Madison, Wisconsin 53711 USA

Library of Congress Catalog Card Number: 81-70161
Standard Book Number: 0-89118-069-9

Printed in the United States of America

Contents

Section IV. Application of Predictions of Soil Physical Properties
and Processes to Prediction of Crop Growth

Foreword

There is an increasing awareness in the U.S.A. and in the world that much of the current level of agricultural production is being achieved at the expense of our nonrenewable soil resources. We can no longer afford to ignore the fact that past and current losses in soil productivity have been largely masked by an increased technological base. This is not to diminish the importance of past technological advances or our need to continue to develop new technology. Rather, we must develop the kind of technology that allows us to at least sustain and hopefully expand our level of agricultural production and at the same time help regenerate rather than deplete our soils.

Reduced tillage systems offer some of the most promising alternatives for reducing soil erosion losses and reducing time and energy requirements for agricultural production. Recognition of the importance of these alternatives has led to expanded tillage research. From this research, it is well documented that alternative tillage systems can reduce soil erosion. However, it is much less clear as to the effect these systems have on soil physical properties and processes.

If alternative tillage systems are to be adopted to the extent needed to effectively control soil erosion it is necessary that we not only be able to measure but also be able to predict their effect on soil physical properties and processes and in turn on crop growth and yield. Today's farmers cannot afford to introduce another major element of uncertainty into their operations.

This publication is the result of a symposium held during the 1980 annual meetings of the American Society of Agronomy and the Soil Science Society of America. The objective of the symposium and the publication is to pursue the goal of predicting the effect of tillage on soil physical properties that are important for plant growth and yield. It has brought together the contributions of some of the most highly qualified scientists in this field today to address problems of great importance to society both today and in the future. We are indebted to the organizer, editors, and authors for this timely and important effort.

C. O. Gardner, ASA President, 1982
R. G. Gast, SSSA President, 1982

ACKNOWLEDGMENTS

The editors are grateful to the organizing committee of the 1980 symposium for their planning and execution. The committee included Dr. R. R. Allmaras, USDA-ARS, Pendleton, Oregon; Dr. D. R. Linden, USDA-ARS, St. Paul, Minn.; Dr. F. D. Whisler, Mississippi State University, Mississippi State, Miss., and Dr. D. M. Van Doren, Jr. (Chair), Ohio Agricultural Research and Development Center, Wooster, Ohio. The other two members of the editorial committee also receive our thanks for their fine efforts; Dr. F. D. Whisler and Dr. E. L. Skidmore, USDA-ARS, Manhattan, Kansas.

Preface

Tillage research has historically been an empirical "science." In a typical tillage experiment, a limited number of tillage tools or systems were compared on a few soils, often using crop growth or yield as the integrator of the environment and sole measured dependent variable. In this way, a wealth of information has been accumulated over the years. Unfortunately, this information is at present difficult to assimilate into a coherent overall pattern.

One reason for the difficulty is the great diversity of weather conditions and soil properties that have differing effects on crop growth. The same comparison among tillage treatments at differing locations may very well have different results, depending upon rainfall pattern, early season soil temperature, soil water holding capacity, soil drainage, or any number of other factors. A second reason is the inability to consistently relate what has been accomplished with tillage to the resulting plant growth and yield factors.

Reliable prediction of the effects of tillage on soil physical properties, and ultimately crop yield, would greatly benefit agricultural advisors or farmers in making management decisions. With a better understanding of the effect of tillage on soil physical properties, probabilities of success with alternative approaches to soil and crop management could be computed on a farm by farm or field by field basis. This would allow us to select the most efficient crop production system for a given situation. Reliable prediction of tillage effects would also greatly reduce the current level of field testing with the attendant plethora of conflicting results.

At the 1980 ASA Annual Meeting, Divisions S-6, S-1, and A-3 sponsored a Symposium entitled "Predicting Tillage Effects on Soil Physical Properties and Processes". The objective of the Symposium and this resultant publication was to demonstrate the potential for achieving the goal of predicting tillage effects on soil physical properties that are important for crop growth and yield. Examples of current progress and problems were presented. These presentations were mixtures of old and new data directed toward a previously untried objective.

With information gained from the Symposium or this publication, persons engaged in planning and executing applied research in tillage and crop production may be encouraged to alter future research to include information helpful for validating various aspects of the prediction process. Persons engaged in graduate education may use the publication to introduce the concepts to their students, whereas those engaged in modeling may be encouraged to attack some of the problems identified during the symposium. Administrators of research programs may wish to encourage these sorts of activities by individuals or groups within their jurisdictions.

P. W. Unger, USDA-ARS, Bushland, Texas-Editor
D. M. Van Doren, Jr., OARDC, Wooster, Ohio-Editor

Chapter 1

Tillage Accomplishments and Potential[1]

W. E. LARSON AND G. J. OSBORNE[2]

The energy crisis, continued excessive erosion on some soils, and the finiteness of our soil resources have renewed our interest in tillage and in farming systems in general, an interest which had lost its urgency following World War II in the USA.

Research and farmers' experience indicate that tillage is responsible for a major part of soil structure deterioration. The adverse effect of tillage on soil structure are well established—oxidation of organic matter by exposure at the surface, mechanical dispersion by puddling through the compaction and shearing action of implements, and by rainfall impact on bare soil. The obvious penalties are soil erosion by wind and water. Less obvious are the reductions in transmission of air and water, both at the soil surface by sealing and at the plow sole. The reductions in air and water movement are less readily observed than the extreme case of impedance to shoot emergence or root penetration, but they can be serious handicaps to crop growth (Pereira, 1975).

Pereira, commenting on the history of tillage in British agriculture quoted from the writings of early essayists such as Virgil, "crude Roman mouldboard ploughs and heavy harrows were followed by the use of mallets to break up the larger clods. The crudeness of the ploughing for weed destruction incurred much subsequent work to pulverise the clods into a seedbed". Comparisons of the accounts of cultivation methods in

[1] Contribution from the Soil and Water Management Research Unit, USDA-ARS, St. Paul, MN, in cooperation with the Minnesota Agric. Exp. Stn., Paper No. 11537. Scientific Journal Series.

[2] Soil Scientist, USDA-SEA-AR, Univ. of Minnesota, and research associate, Univ. of Minnesota, St. Paul, MN 55108.

Fitzherberts' Boke of Husbandry in 1523 with that of Virgil's indicates that apart from the reinforcement of the wooden moldboard plow with an iron plowshare there had been no effective advance in tillage in 15 centuries (Pereira, 1975). Adherence to intensive land preparation systems has resulted in severe soil erosion on much of the American continent. Wind and water erosion is excessive on approximately one-third of the cropland in the USA. Williams (1967) reported that an estimated 4 billion tons of sediment enter surface waters in the USA annually. For every bushel of corn produced in Iowa it is estimated that the equivalent of 2 bushels of soil are lost. These are figures that must be considered when the economics of long-term cropping are being assessed.

EROSION AND TILLAGE

Effectiveness of tillage systems in minimizing soil erosion depends on soil and topographic conditions. Lindstrom et al. (1979) calculated the average erosion rate for all cultivated soils in the Corn Belt when different tillage practices were used. For conventional tillage (fall moldboard, disc, plant) the average erosion was 21.5 metric tons ha^{-1} year^{-1}; for chisel-plow (3,920 kg ha^{-1} of residue on the soil surface) the average erosion was 8.7 metric tons ha^{-1} year^{-1}; and for no-tillage (3,920 kg ha^{-1} of residue on the soil surface) the average erosion was 6.5 metric tons ha^{-1} year^{-1}. Since the average soil loss tolerance (T) is 9 metric tons ha^{-1} year^{-1}, one might conclude that if all corn (Zea mays L.) and soybeans [Glycine max (L.)] were grown with no-tillage or conservation tillage, erosion could be controlled. However, only on two of the six Major Land Resource Areas of Iowa and Minnesota would improve tillage alone reduce the average soil loss below T (Onstad et al., 1981).

Use of conservation or no-till would significantly reduce soil loss on all Major Land Resource Areas of the Southeast, although it would not bring erosion below the tolerance level on most of them (Campbell et al., 1979).

Skidmore et al. (1979) calculated that wind erosion could be controlled on 55% of the cropland in the Great Plains if a tillage system were used that left all residues on the surface and the surface was smooth. If the soil surface was rough, wind erosion could be controlled on 87% of the land if all residues were maintained on the soil surface.

Conservation tillage practices that leave crop residues on the soil surface can also increase water infiltration into the soil. Onstad and Otterby (1979) estimated that conservation tillage could increase retained water for straight-row corn on soils with moderate infiltration rates from 0.5 cm (0.2 inches) in the Great Plains to 5.0 cm (2 inches) in the Southeast. On soils with slow infiltration rates, the increase would range from 2.5 cm (1 inches) in the Great Plains to 12.5 cm (5 inches) in the Southeast for conservation tillage. According to these estimates, runoff would be eliminated for most small storms and reduced for all storms. This increased soil water storage may have a material impact on crop yields.

ENERGY USED IN TILLAGE

Modern agriculture in North America, Europe, and elsewhere is energy intensive in terms of liquid fuel consumption. As energy input has increased, labor input has decreased (Fig. 1). For example, the American farmer spent 150 min producing 25 kg (1 bu) of corn in the early 20th century and about 61 min in 1955. Today, he spends less than 3 min per 25 kg (1 bu) (Hayes, 1976).

From a review of the literature, Crosson[3] concluded that no-till saves 28 to 37 liters ha^{-1} (3 to 4 gal/acre) of diesel fuel and other forms of conservation tillage save 9 to 28 liter ha^{-1} (1 to 3 gal/acre) as compared with conventional tillage.

About 2.5% of the total energy consumed in the USA is used in agriculture; of this 2.5%, tillage uses about 5%. The major areas of energy consumption in crop production are: fertilizers, 33%; grain drying, 16%; irrigation, 13%; and pesticides, 5%. Other significant uses of energy are: harvesting, transportation, frost protection, and product handling. Even though tillage accounts for a very small percentage of the total USA

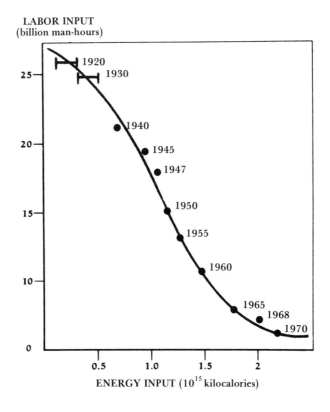

Fig. 1. The substitution of energy for labor on U.S. farms.

Table 1. Total energy for tillage, planting and weed control (adapted from Griffith et al., 1977).

| Tillage system | Fuel | Energy requirement | | | Diesel fuel equiv. |
		Indirect mach.[†]	Weed control	Total	
		kcal × 10³ ha¹			liters
Conventional (Broadcast herb.)	320	160	311	791	85.6
Conventional (Band herb.)	130	160	147	648	70.1
Chisel	228	114	342	698	75.5
Coulter (No-till)	91	46	408	596	64.5

[†] Assumes that the energy used in machinery manufacture is one-half that of fuel consumption.

energy consumption, there is a potential for savings from improved tillage practices. Savings from direct fuel consumption, or decreased use of fertilizers and pesticides, and could result from the selection of the best tillage practice.

The total energy used in three tillage systems for corn on four Indiana soils is given in Table 1 (Griffith et al., 1977). Fuel used for conventional (spring plow), chisel, and coulter (no-till) was 320, 228, and 91 × 10³ kcal ha⁻¹, respectively. The total energy used for conventional (broadcast herbicide), conventional (band herbicide), chisel, and coulter (no-till) was 791, 627, 684, and 545 × 10³ kcal ha⁻¹, respectively. The total saving in fuel equivalents as compared to conventional (broadcast herbicide) was 15, 10, 21 liters ha⁻¹ for conventional (banded herbicide), chisel, and coulter (no-till), respectively.

Phillips et al. (1980) calculated that a 46% savings in energy consumption could be realized from no-tillage as compared with conventional tillage for corn (728 vs. 395 × 10³ kcal). Greater energy was consumed in conventional tillage as compared with no-tillage for machinery manufacturing and repair. More energy was consumed in no-tillage for herbicides and insecticides. Phillips et al. (1980) estimated that the savings in diesel fuel for tractors to power tillage equipment was about 33 liters ha⁻¹ for no-tillage corn as compared with conventional and about 31 liters ha⁻¹ for soybeans.

Soil compaction from previous tillage and wheel-traffic can have a measurable effect on the energy required for tillage the following year. Voorhees (1980) found that diesel fuel consumption during moldboard plowing increased from 25.6 to 34.6 liters ha⁻¹ when the previous passes by a tractor increased from 0 to 5 on a Nicollet clay loam (fine-loamy, mixed mesic Aquic Hapludolls).

CROP YIELDS

Crops respond to changes in soil water content, soil temperature, nutrient supply, composition of the soil atmosphere, and to the strength of the soil. The specific tillage practice employed influences all these plant growth factors, although the effects may be different in different soils and weather conditions. The specific response to a soil physical change may depend on the plants' physiological growth stage.

Van Doren and Triplett (1969) examined the results of experiments where corn growth after emergence in tilled (plowed plus secondary tillage) soil was compared with growth in non-tilled soil. The data used for their comparisons had equal plant populations and weed control. Their findings taken from research in Ohio and Virginia, are of particular interest. No-tillage planting of corn following a row crop in clay loam to clay soils produced lower yields than the fall-plowed conventional tillage system. Corn yield from the two tillage systems were equal on the clay loam to clay soils following sod and on the silt loam soils following a row crop. No-tillage planting of corn following sod on silt loam soils produced substantially greater yields compared with the spring plowed conventional tillage systems in both states. They concluded that "this apparent interaction between soil type and previous crop should be examined to establish major causes for variations in yield differences between tillage treatments".

Phillips et al. (1980) report that "except for a few unusual situations, soil water content is almost always higher under the no-tillage system than under conventional tillage." This is attributed to reduction of evaporation losses due to the mulch on the surface.

There is considerable evidence, however, that more continuous macropore systems are developed under no-till. Tillage which shears the soil at some depth below the surface, seals off channels developed by plant roots, or shrinkage cracks which conduct water to lower levels for storage in, or drainage from the profile. Tillage tends to increase the soil water levels in the plowed layer which leads to increased evaporation losses (Wittmus and Yazar, 1980). From 4 years of observations, Ehlers and van der Ploeg (1976) noted that at water potentials of -100 mb or greater, hydraulic conductivity was higher in untilled than tilled soil. They concluded that larger pores are broken up in tilled soil but remain continuous in untilled soil.

Greb et al. (1970) observed that increased amounts of soil water storage occurred under increasing depths of mulch with stubble mulch tillage. Unger et al. (1971) working on a clay loam soil found that cultivation without herbicide limited profile water additions to the upper 75 cm while herbicide treatment with or without cultivation resulted in profile water additions down to about 120 cm.

Cannell et al. (1978) suggested that the most common soil problems under English conditions (mulch burned or removed) giving rise to yield reductions under no-till were soil compaction often with associated waterlogging and lack of surface tilth. These authors pointed out, however, that on well-structured soils, especially on some clay soils, these changes are poor indices of the suitability of the soil for root growth. In particular, on such soils which have been no-tilled for 2 or 3 years, there is evidence of changes in soil conditions some of which may lead to improvements in root growth. For example, Ellis et al. (1979), working on a clay soil in England, showed that within 5 weeks after planting in each of 4 years that no-tilled soil was more compact as measured both by bulk density and by penetrometer resistance. On this site there was no evidence of restricted root growth during early seedling development similar to that which occurred with spring barley on a sandy-loam soil (Ellis et al., 1977). Field observations by the above authors and the results of Osborne (1981) for porosities and conductivities (air and water) suggest that the lack of the expected relationship between bulk soil properties and root growth may be due to the markedly greater continuity of cracks in clay soils when they are not disturbed. These higher air porosities may explain the higher observed concentrations of oxygen and the lower water contents in the 10 to 20 cm depth in these soils during the winter under no-till (Dowdell et al., 1979).

Griffith et al. (1973) in Indiana and Olson and Schoeberl (1970) in South Dakota found that, of the systems tested, the till plant systems gave highest yield of corn. In Indiana, till planting was on a ridge. Moldenhauer (1976) considers that more favorable early season temperatures resulting from planting on the ridge may have been responsible for the higher yields from till planting compared to the other tillage systems. Griffith et al. (1973) states that "in general, as amount of tillage decreased and ground cover increased, plant growth was slowed and maturity was delayed in northern and eastern Indiana soils". They also state that, "till planting may also be competitive on fine-textured, poorly drained soils if used in conjunction with pronounced residue-free ridges to achieve better drainage, thus improving warming and drying". Behn (1973) reported success on the poorly drained, Webster, silty clay loam soil of Iowa using residue-free ridges and planting with a till-planter.

To summarize for the USA, corn yields on well-drained soils appear to be about the same with conservation (including no-till) as with conventional tillage. On coarse-textured soils and soils with low water-holding capacity, yields may be higher from tillage practices that leave residues on the surface. On less-well-drained and poorly drained soils, however, such practices may decrease crop yields. Other observations for the Eastern Corn Belt as outlined by Griffith et al. (1977) are: (a) shallow tillage and no-tillage for corn are better suited to poorly drained soils when corn follows anything but corn; (b) corn on poorly structured soils low in organic matter is likely to react positively to surface residue tillage, because of reductions in crusting and water runoff; and (c) surface residue tillage systems are better adapted to the longer and warmer growing seasons in the southern half of the Corn Belt and further south. The above generalizations assume that equal plant populations are obtained for all tillage practices. However, frequently plant populations are lower from the various forms of conservation tillage as compared to conventional tillage (Griffith et al., 1977). Improved planters are now on the market which alleviate or eliminate this problem. Weed control also requires modifications of accepted procedures and as with any farming system, if weeds are not controlled, yields may be reduced.

FUTURE CHANGE IN TILLAGE PRACTICES

The use of no-tillage or conservation (reduced) tillage is increasing. Crossón[3], based on Soil Conservation Service data, estimates that the percentage of harvested cropland in conservation tillage has steadily in-

---

[3] Crosson, Pierre. 1980. Conservation tillage: An assessment (Unpublished manuscript).

creased from 2.3 in 1965 to 16.1 in 1979 (Fig. 2). Over 20% of the harvested cropland is in conservation tillage (including no-till) in the Northern Plains, Southeast, Appalachia, Cornbelt, and Mountain regions.

In Kentucky, about 20% of the corn and soybean area was no-tilled in 1978. In Iowa, about 50% of the harvested corn and soybean area was not moldboard plowed in 1978 (Soil Conservation service data). Most of the area not moldboard plowed was either chisel plowed or disked. Less than 1% was no-tilled.

While reduced tillage systems are our main defense against wind erosion, they are used on only about one-third of the susceptible area.

The percentage of tillable land suitable for conservation tillage in Ohio, Indiana, Illinois, and Iowa as well as the percentage of land in conservation tillage in 1979 has been reported by Crosson[3] (Table 2). The amount of land suitable for conservation tillage (taken from Cosper, 1979) is based on the assumption, supported by experimental results, that soil with slow internal drainage and in areas of higher rainfall is less suited for conservation tillage. The projections in Table 2 indicate that both the percentages of land suitable for and now in conservation tillage increase as one moves westward from Ohio to Iowa.

Crosson[3] estimates that conservation tillage will be used on 50 to 60% of the nation's cropland by the year 2010. This is a more conservative estimate than others have given. Iowa State University's model of U.S. Agriculture using projections of production, crop yields, and soil erosion by Crosson[3], indicates 75% of the cropland might be in conserva-

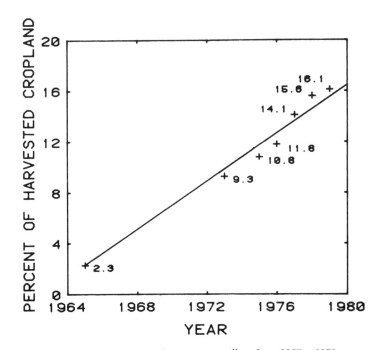

Fig. 2. Increase in use of conservation tillage from 1965 to 1979.

Table 2. Percentages of land apt for conservation tillage and in conservation tillage, Ohio, Indiana, Illinois, and Iowa (from Crosson[3]).

State	Percent apt for conservation tillage[†]	Percent in conservation tillage, 1977[‡]
		%
Ohio	47.5	8.0
Indiana	53.4	22.8
Illinois	65.9	28.0
Iowa	76.4	38.9

[†] Cosper (1979).
[‡] Lessiter (1979).

tion tillage by 2010. The USDA (1975) in a preliminary technology assessment of minimum (conservation) tillage estimated that more than 80% of the cultivated land could be in conservation tillage by the year 2000, and nearly one-half of all crop area could be no-tilled by that time.

Crosson's[3] estimate of the cropland area suitable for conservation tillage was based on the premise that crop yields would be equal to, or higher than, those obtained using conventional tillage. As pointed out earlier in this article, significant amounts of energy can be saved and erosion can be materially reduced by the use of conservation tillage. In view of these seemingly major advantages of conservation tillage, we must rapidly increase the cropland area in conservation tillage.

MODELING

Tillage research has historically been an empirical "science". A large volume of information has been accumulated over the years, which is at present difficult to assimilate into an overall pattern so that site specific results might be projected over a broad area. A difficulty in generalizing from tillage information is our inability to consistently relate the soil changes accomplished by tillage (soil water content, soil temperature, soil aeration, and soil strength) to the resulting plant growth.

Reliable prediction of tillage effects on soil physical properties, and ultimately crop yield, will: (a) greatly benefit agricultural advisors or farmers in making management decisions, (b) allow selection of the most efficient production system for a given soil and climate, and (c) reduce the current level of field testing or change the emphasis of tillage studies.

Because of the foregoing, we need an accurate, site specific, and rapid information delivery system that can be quickly updated as weather, cropping plans, or economic conditions change. Because of the complexity and dynamic nature of the system and the large number of variables to be considered, we need to organize, in a systematic and quantitative way, what we know about the tillage needs of a soil for optimum crop production and erosion control.

Crop growth and development models that are based on physiological, phenological, and physical principles and controlled by climatic inputs enable quantitative description of the dynamic crop production system (Stapper and Arkin, 1979). While the degree of sophistication (or complexity) of these models has increased over the past 20 years (deWit, 1958; Dale and Shaw, 1965; Baier and Robertson, 1966; Saxton et al., 1974; Childs et al., 1977) the effects of tillage on the optimization of soil and water resources has not been interfaced (incorporated) in these models.

Erosion models have been developed for designing erosion control systems, predicting sediment yield for reservoir design, predicting sediment transport, and simulating water quality. Also, soil characteristics have been used to compute soil productivity ratings. However, erosion models have not been linked with crop growth models to form the necessary structure to study the erosion—productivity problem (Williams et al., 1981).

National modeling teams in the USA are working on closely related problems-crop growth and non-point source pollution. The non-point-source pollution team had developed a field-scaled chemical transport model called CREAMS which does not consider tillage or crop growth, while the crop-growth teams are developing plant yield models with particular emphasis on economically important crops such as cotton (*Gossypium hirsutum* L.), wheat (*Triticum aestivum* L.), corn, and soybeans (Williams et al., 1981).

In St. Paul, Minn. we are working on a crop production model currently referred to as teh NTRM (nitrogen-tillage-residue management), to be used in agricultural research with so-called user models being developed from the model for use by advisory services and for direct access by farmers (Shaffer et al., 1980).

Modeling is a powerful tool that is well suited for soil tillage research. In addition to the goal of providing accurate, rapid, site specific projections of tillage management and crop yeidls, a model can provide: (a) an analytical mechanism for the scientist to study the system and (b) a communication tool for disseminating information between scientists and to the public.

We should begin to organize the wealth of knowledge about tillage into a systematic information network that will aid researchers in determining research directions, as well as farmers in making crucial production decisions.

We are extremely fortunate to be involved in tillage research at this time. Never before have we had the modeling techniques and computers available to attack such a complex problems as soil tillage. The next few years will be a challenging time.

LITERATURE CITED

1. Baier, W., G. W. Robertson. 1966. A new versatile soil moisture budget. Can. J. Plant Sci. 46:299–315.
2. Behn, E. E. 1973. Acceptance of conservation tillage. Role of the farmer. *In* Conservation tillage. Proc. of a National Conf. Des Moines, Iowa, 28–30 Mar. 1973. Soil Conserv. Soc. Am., Ankeny, Iowa.

3. Campbell, R. B., T. A. Matheny, P. G. Hunt, and S. C. Gupta. 1979. Crop residue requirements for water erosion control in six southern states. J. Soil Water Conserv. 34: 83–85.

4. Cannell, R. Q., D. B. Davies, D. Mackney, and J. D. Pidgeon. 1978. The suitability of soils for sequential direct drilling of combine-harvested crops in Britian: a provisional classification. Outlook Agric. 9:306–316.

5. Childs, S. W., J. R. Gilley, and W. E. Splinter. 1977. A simplified model of corn growth under moisture stress. Am. Soc. Agric. Eng. Trans. 20:858–865.

6. Cosper, H. R. 1979. Soil taxonomy as a guide to economic feasibility of soil tillage systems in reducing nonpoint pollution. p. 26. Economics, Statistics, and Cooperatives Service USDA. Staff Report.

7. Dale, R. F., and R. H. Shaw. 1965. The climatology of soil moisture, atmospheric evaporative demand, and resulting moisture stress days for corn at Ames, Iowa. J. Appl. Meteorol. 4:661–669.

8. deWit, C. T. 1958. Transpiration and crop yields. Institute of Biological and Chemical Research on Field Crops and Herbage, Wageningen, the Netherlands, Verse-Land-bouwk, onder 2, No. 64, 6S Gravenhage.

9. Dowdell, R. J., R. Cress, J. R. Burford, and R. Q. Cannell. 1979. Oxygen concentrations in a clay soil after ploughing or direct drilling. J. Soil Sci. 30:239–245.

10. Ehlers, W., and R. R. van der Ploeg. 1976. Evaporation, drainage and unsaturated hydraulic conductivity of tilled and untilled fallow soil. Z. Pflanzenernaehr. Bodenkd. 3: 373–386.

11. Ellis, F. B., J. G. Elliott, B. T. Barnes, and K. R. Howse. 1977. Comparison of direct drilling, reduced cultivation and ploughing on the growth of cereals. 2. Spring barley on a sandy loam soil:soil physical conditions and root growth. J. Agric. Sci. 89:631–642.

12. ————, ————, F. Pollard, R. Q. Cannell, and B. T. Barnes. 1979. Comparison of direct drilling, reduced cultivation and ploughing on the growth of cereals. J. Agric. Sci. 93:391–401.

13. Greb, B. W., D. E. Smika, and A. L. Black. 1970. Water conservation with stubble mulch fallow. J. Soil Water Conserv. 25:58–62.

14. Griffith, D. R., J. V. Mannering, H. M. Galloway, S. D. Parsons, and C. B. Richey. 1973. Effect of eight tillage-planting systems on soil temperature, percent stand, plant growth, and yield of corn on five Indiana soils. Agron. J. 65:321–326.

15. ————, ————, and C. B. Richey. 1977. Energy requirements and areas of adaptation for tillage-planting systems for corn. In William Lockeretz (ed.) Agriculture and energy. Academic Press, New York.

16. Hayes, Denis. 1976. Energy: The case for conservation. Worldwatch Paper 4. Worldwatch Institute, Washington, DC.

17. Lessiter, Frank. 1979. No-till moving ahead. p. 4–5. In No-till farmer.

18. Lindstrom, J. J., S. C. Gupta, C. A. Onstad, W. E. Larson, and R. F. Holt. 1979. Tillage and crop residue effects on soil erosion in the Corn Belt. J. Soil Water Conserv. 34: 80–82.

19. Moldenhauer, W. C. 1976. Tillage systems. In W. E. Larson (ed.) Conservation tillage research progress and needs. ARS-NC-57. USDA.

20. Olson, T. C., and L. S. Schoeberl. 1970. Corn yields, soil temperature, and water use with four tillage methods in the western Corn Belt. Agron. J. 62:229–232.

21. Onstad, C. A., W. E. Larson, S. C. Gupta, and R. F. Holt. 1981. Maximizing crop residues for removal in Iowa and southern Minnesota. J. Environ. Qual. (In press).

22. ————, and M. A. Otterby. 1979. Crop residue effects on runoff. J. Soil Water Conserv. 34:94–96.

23. Osborne, G. J. 1981. Soil structure and farming with minimum cultivation. p. 84–94. In J. Logan (ed.) Proc. for the national Spray-Seed Conf. 1981. Albury, New South Wales, Australia. Imperial Chemical Industries, Melbourne, Australia.

24. Pereira, H. C. 1975. Agricultural science and the traditions of tillage. Outlook Agric. 8:211–212.

25. Phillips, R. E., R. L. Blevins, G. W. Thomas, W. W. Frye, and S. H. Phillips. 1980. No-tillage agriculture. Science 208:1108–1113.

26. Saxton, K. E., H. P. Johnson, and R. H. Shaw. 1974. Modeling evapo-transpiration and soil moisture. Am. Soc. Agric. Eng. Trans. 17:673–677.

27. Shaffer, M. J., S. C. Gupta, J. A. E. Molina, D. R. Linden, C. E. Clapp, and W. E. Larson. 1980. Simulating crop response to tillage: an integrated approach. p. 15. *In* Agron. Abstr., Am. Soc. of Agron., Madison, Wis.

28. Skidmore, E. F., M. Kumar, and W. E. Larson. 1979. Crop residue management for wind erosion control in the Great Plains. J. Soil Water Conserv. 34:90–94.

29. Stapper, M., and G. F. Arkin. 1979. Simulating maize dry matter accumulation and yield components. Winter meeting, ASAE, New Orleans, La. Paper No. 79-4513.

30. Steinhart, C. E., and J. S. Steinhart. 1974. Energy: Sources, use, and role in human affairs. North Scituate, Mass.

31. Unger, P. W., R. R. Allen, and A. F. Wiese. 1971. Tillage and herbicides for surface residue maintenance, weed control, and water conservation. J. Soil Water Conserv. 26:147–150.

32. U.S. Department of Agriculture, Office of Planning and Evaluation. 1975. Minimum tillage: A preliminary assessment.

33. Van Doren, D. M., Jr., and G. B. Triplett, Jr. 1969. Mechanism of corn (*Zea mays* L.) response to cropping practices without tillage. Research Circular 169, Ohio Agricultural Research and Development Center, Wooster, Ohio.

34. Voorhees, W. B. 1980. Energy aspects of controlled wheel traffic in the northern Corn Belt of the United States. p. 333–338. *In* Vol. 2, Proc. Int. Soil Tillage Research Organization, 8th Conf. 1979. Univ. of Hohenheim, Germany.

35. Williams, D. A. 1967. Tillage as a conservation tool. p. 56–57, 70. *In* Tillage for greater crop production. Am. Soc. Agric. Eng. St. Joseph, Mich.

36. Williams, J. R., R. R. Allmaras, K. G. Renard, Leon Lyles, W. E. Moldenhauer, G. W. Landgale, L. D. Meyer, W. J. Rawls, R. Daniels, and R. Magleby. 1981. Soil erosion effects on soil productivity: A research perspective. J. Soil Water Conserv. 36:82–90.

37. Wittmuss, Howard, and Attila Yazar. 1980. Moisture storage, water use and corn yield for seven tillage systems under water stress. p. 66–75. *In* Proc. Crop Production with Conservation in the 80's. ASAE Seminar. 1–2 December, Chicago, Ill. (ASAE 7-81).

Chapter 2

Changing Soil Condition—The Soil Dynamics of Tillage[1]

ROBERT L. SCHAFER AND
CLARENCE E. JOHNSON[2]

ABSTRACT

Any manipulation that changes soil condition may be considered as tillage. Most often machines are used to apply forces to the soil to effect this change. Soil dynamics is a description of the behavioral response of soil to applied forces and of soil-machine behavior. The state of development of soil dynamics—quantitative descriptions of soil behavior, soil-machine behavior, and resultant soil condition—is explored. Research needs and directions in soil dynamics related to tillage and to prediction of the resultant soil condition are discussed.

Any manipulation that changes soil condition may be considered tillage. This includes tillage for such purposes as weed control and incorporation of soil amendments. The art of tillage began when man first domesticated and cultivated plants. Man observed plant responses to certain soil manipulations. Tillage tools have evolved from rudimentary ones operated by humans to more sophisticated ones powered by animals and, eventually, by machines. Tillage began as a science when man attempted

[1] Contribution from National Tillage Machinery Lab., USDA-ARS, in cooperation with Auburn Univ. and Alabama Agric. Exp. Stn., Auburn, AL.

[2] Director, National Tillage Machinery Lab., USDA-ARS-AR, Auburn, AL 36830; and professor, Agricultural Engineering Dep., Auburn Univ. and Alabama Agric. Exp. Stn., Auburn University, AL 36849.

to describe and quantify the soil condition that improved plant growth. As soil manipulations were perceived, tools were developed. For example, when there was a need to invert or pulverize soil, plows were developed, and when there was a need to sever weeds, cultivators were developed. The search for effective and efficient tillage tools led to investigations of soil response to applied forces and to investigations of soil-machine behavior. Thus, the development of soil dynamics began.

We were challenged to address the general area of tillage effects on soil physical properties and processes. The specific questions we were asked to address were:

1) What is the effect of a specific tillage tool or vehicle on a specific soil?
2) What are the best (if any) currently available predictive (models) equations for these effects?
3) What information is required to make the predictions?
4) What recommendations do you have for improving the predictions?

Scientific investigation of tillage should provide improved answers to these questions and advance tillage from an art to a science. We will limit our comments to tillage; others are better qualified to discuss vehicle effects on soil.

Conceptually, tillage tools apply forces to soil which causes soil motion that changes the soil condition for enhanced agricultural production; e.g., by increasing emergence, improving plant rooting, increasing infiltration, and controlling erosion. In addressing questions 1 to 4, we will raise additional questions concerning the physical behavior of soil in response to tillage forces.

The active and passive behavioral response of soil to forces applied by machine, plants, and the environment influences the results of tillage. The investigation of soil behavior in response to applied forces and of soil-machine behavior was the start of the development of soil dynamics. One may raise the questions, "What is soil dynamics?", "How does soil dynamics relate to tillage?", and, more explicitly, "Is an understanding of soil dynamics pertinent to answering questions 1 to 4?"

Soil dynamics may be defined as the relation between forces, soil deformation, and soil in motion. This definition does not restrict the type of force system or the purpose for applying the force system. However, in this paper our primary interest is the application of mechanical forces by machines to change the soil condition for agricultural production purposes.

For a framework for discussion and a prospectus for relating soil dynamics to tillage, we will use an analogy. Consider the body of knowledge that defines aerodynamics—a segment of fluid mechanics and thermodynamics—which has greatly influenced developments in automotive design, aircraft design, and space travel.

There is a contrast in the complexity of the medium air, in aerodynamics, compared to the medium soil, in soil dynamics. An airfoil moves through air and a tillage tool moves through soil, but air is a much more continuous medium than soil. Furthermore, air can be considered a homogeneous and isotropic mixture of particles, whose sizes are very much smaller than an airfoil moving through them, in contrast to soil which may contain aggregates, clods, and foreign material whose sizes are nearer to the size of the tillage tool. In contrast to air, soil is non-homogeneous and often exhibits anisotropic behavior. In addition, tillage forms discontinuities (shear planes) within the soil. Thus, a description of soil behavior must necessarily be much more complex than a description of air behavior.

Air travel has advanced from an art in the Wright brothers' era to a science in our present space travel. Aerodynamics has been a key element in that progress during a time span of less than 100 years.

Soil dynamics could have a similar impact on tillage; unfortunately, the state of knowledge in soil dynamics is not as advanced as in aerodynamics, nor has the rate of knowledge increase been the same. However, much less scientific manpower has been devoted to the development of soil dynamics than to the development of aerodynamics; perhaps, because of the complexity of soil behavior. Interestingly, a leading scientist in soil dynamics and tillage—Walter Soehne—was trained in aerodynamics.

THE ROLE OF SOIL DYNAMICS

As a basis for further exploring the relation of soil dynamics to tillage, in this section the authors discuss the concepts of behavioral properties and state properties. State properties describe a material without regard to intended use. As an example, a wire may be characterized by its chemical composition, density, and color; these are state properties. On the other hand, behavioral properties describe the reaction of a material to an applied force system. For example, if a voltage is applied across a wire, the amount of current flowing through the wire depends on the resistance of the wire (Ohms Law). Resistance is a behavioral property. Also, if the wire is stretched by force applied to its ends, the amount of deflection depends on the modulus of elasticity (Hookes Law). The modulus of elasticity is a behavioral property. Both of these completely different behaviors—current flow and deflection—are important in the description of the wire based on its intended use, but they must be described separately. Further, although it may be possible to relate behavior of the wire to the state properties, the state properties may not be rationally descriptive of the wire's behavior.

When a tillage tool is used to apply forces to soil, the soil moves and its condition changes. Behavioral properties must be used to describe its action. Unfortunately, in the past, state properties, particularly moisture content and density, have often been primary parameters in describing tillage behavior. However, unless the relations between behavioral and state properties are unique and are known, the use of state properties to describe the dynamic tillage action is not a rational approach. State properties have probably been used because they are more obvious and more easily quantifiable than behavioral properties. Behavioral properties often are very difficult to quantify, but we must undertake and complete that task.

In agriculture, we apply active force systems—tillage—to prepare seedbeds and rootbeds, incorporate amendments, control weeds, control pests, enhance infiltration, and control erosion. The state of the soil is changed from its initial condition to some final condition as the result of the applied forces and of the resulting soil movement. Cooper and Gill (1966) illustrated that idea with the conceptual relation:

$$S_f = f(S_i, F),$$ [1]

where
  S_f = final soil condition
  S_i = initial soil condition
  F = mechanical forces applied to the soil.

They expressed another simplified conceptual relation that may be of more interest from a production standpoint:

$$CP = g(S, E, P, M),$$ [2]

where
  CP = crop production
  S = soil composition and condition
  E = environment
  P = plant species
  M = management practices.

The conceptual relations in Eq. [1] and [2] were stated in a very simplified manner without mathematical rigor. This is analogous to a scientist selecting pertinent factors and then developing a factorial statistical design to explore main effects and interactions in data (Steel and Torrie, 1960). Development of those two conceptual relations into mathematical equations for predictive purposes is the real crux of our challenge—answers to questions 1 to 4.

Gill and Vanden Berg (1967) and Vanden Berg and Reaves (1966) expressed two additional generalized relations which reflect aspects of the tillage machine system:

$$F = h(T_s, T_m, S_i)$$ [3]

and

$$S_f = k(T_s, T_m, S_i),$$ [4]

where
  T_s = tool shape
  T_m = manner of tool movement.

They referred to these abstract relations as the force tillage equation (Eq. [3]) and the soil-condition equation (Eq. [4]). Much of the past research on soil-machine relations and soil dynamics has been related to the concepts of Eq. [3]—the force tillage equation.

A change from S_i to S_f is caused by soil movement. This change involves strain and yield of the soil; so, force-movement relations of soil are important when soil condition is changed. Some research has been related to the concepts of Eq. [1]—the basic processes of soil deformation, e.g., stress-strain behavior. Such research has been concerned with soil stress, stress distribution, strain, strain distribution, soil strength, soil yield (shear, compression, tension, and plastic flow), and rigid body movement (momentum, friction, adhesion, and abrasion).

Several different quantities or soil properties may be required for adequate quantification of each of the abstract entities, S, S_i, and S_f. S_i in Eq. [3] must be quantified in terms of the soil's resistance to deformation and movement, whereas S, S_i, and S_f in Eq. [2] and [4] must be quantified in terms of the soil's strength and of its resistance to water, air, and heat flow.

Most objectives of past research have not been aimed at empirical or theoretical definition of the concepts of Eq. [1] to [4]. Rather, they have been aimed at relating the differential change in the "dependent factor" as influenced by a change in one or more "independent factors." That is, differential change in S_f, CP, and F have been studied with respect to changes in individual "independent factors" or in combinations of "independent factors." Conceptually, the differential change in a "dependent factor" with respect to an "independent factor" may be (1) a constant, (2) a function of that independent factor, (3) a function of one or more of the other independent factors, or (4) a function of that independent factor and one or more of the other independent factors. Cases 3 and 4 are interactions, as defined in statistical analyses (Steel and Torrie, 1960). Interactions increase the difficulty of developing empirical relations.

With respect to Eq. [1], work by Dunlap and Weber (1971) and Kumar and Weber (1974) suggests complicated interactions. They found that the final soil condition has some dependence on the stress path (stress history) of the applied load. Their results indicated that Eq. [1] and [3] may be more complex than behavioral relations in other technologies— say, aerodynamics. However, their results suggested that the energy efficiency of one force system applied to create a final soil condition may differ from that of another force system applied to create the same final soil condition. So, the energy efficiency of the tillage process depends on how the tillage machinery applies force to the soil. Soil dynamics involves defining Eq. [1], [3], and [4] in rigorous mathematical terms, rather than conceptually, to provide some fundamental foundations for Eq. [2].

SOIL CONDITION CHANGE

Soil Behavior

When a soil is tilled, it is changed from S_i to S_f because it yields, fails, and moves as influenced by the force system, F (Eq. [1]). Soil yields and fails when its strength is overcome. Four types of soil failure in terms of stress-strain behavior have been observed: shear, compression, tension, and plastic flow. A tillage tool may apply a force system that creates all four types of failure. The type and extent of each type of failure caused by tillage determine the final soil condition, S_f.

Soil strength is commonly defined in the context of the four types of failure. Since agricultural soils vary from near-liquid to very brittle materials, soil strength and soil failure are often complex and confusing entities.

Shear behavior is defined and discussed in many textbooks, e.g., Yong and Warkentin (1966). Different devices (e.g., grousered annulus, direct shear box, vane, or cone penetrometer) are often used to quantify shear behavior. However, the results often depend on the devices (Bailey and Weber, 1965; Dunlap et al., 1966). The triaxial shear test is well accepted, but it is not practical to use for the wide range of conditions found in agriculture.

Failure of soil by compression is generally associated with volume change. Researchers are still searching for adequate stress-volume-strain relations for agricultural soils that are not influenced by soil type. Some measurement methods give misleading results because other types of failure are present.

Tension failure has the same meaning in soils as in other materials. Tension failure occurs when complete separation occurs. Techniques have been devised for quantifying tension failure (Hendrick and Vanden Berg, 1961). However, the role of tension failure in tillage has not been clearly established.

Plastic flow has been observed in soils, particularly clayey soils. However, it has never been clearly quantified in terms of stress-strain behavior, other than conceptually. The plastic limit—one of the Atterburg limits (Yong and Warkentin, 1966)—gives a measure of a soil's consistency. The plastic limit, influenced by texture, is usually between 15 and 70% moisture content (dry weight basis). Plastic flow is said to occur when a subsoil tillage tool moves through wet clay; the soil may move around the subsoiler as a continuous mass with no observed separation.

In addition to causing failure, the tillage tool also moves the soil after it has failed. Portions of soil (clods and aggregates) may move as rigid bodies along the tool surface or within the soil mass. Friction, adhesion, and momentum describe this rigid body movement.

Momentum is the product of mass and velocity. The force system applied by a tillage tool changes the soil's momentum during its contact with the tillage tool. When one rigid body of soil in motion contacts other soil with a different motion, an impulse of force is created that may cause further soil failure. The Newtonian laws of motion rigorously represent rigid body soil behavior.

Frictional forces are generated when a soil mass moves relative to and in contact with another material (tillage tool surface) or another soil mass. The Coulomb friction concept, commonly described in physics and engineering textbooks (e.g., Higdon and Stiles, 1957), seems to adequately describe this behavior. Coulomb friction is mathematically defined as

$$\mu = F_f/N = \tan \psi, \qquad [5]$$

where

μ = coefficient of friction
F_f = frictional force tangent to the surfaces
N = normal force perpendicular to the surfaces
ψ = friction angle whose tangent is μ.

The μ value varies greatly with soil type, soil condition, and type of tillage-tool surface material. Typically, the μ value for soil on steel ranges from 0.2 to 0.7. At present, μ must be measured for each situation of interest since a predictive model relating it to such factors as soil type, soil condition, and tillage-tool surface material has not been developed.

Adhesion is the tension force required at the mutual contact surface of two rigid bodies to separate them. Nichols (1931) presented a comprehensive discussion of soil-on-material sliding that included a discussion of friction and adhesion. The concepts of adhesion are well developed. Adhesion of soil to the tillage tool causes a normal load on the mutual-contact surface. Since frictional forces are a function of the normal load, adhesion adds to the frictional forces. The problem then is measuring adhesion and μ jointly. The conventional method for such measurement is to slide the tillage-tool material on soil at various external normal loads. The following equation adequately represents this phenomenon.

$$S = A + \sigma \tan \psi, \qquad [6]$$

where
    S = tangential stress at the soil-metal interface,
    A = apparent soil-material adhesion
    σ = normal stress on the soil-material interface
    ψ = apparent angle of soil-material friction.

Like friction, soil-material adhesion varies greatly with soil type and soil condition; thus, adhesion must be measured for each situation of interest.

The tangential stress at the tillage-tool surface greatly influences the final soil condition, S_f. This influence has been noted by several researchers. The plastic-covered moldboard (Cooper and McCreery, 1961) and lubricated plow (Schafer et al., 1975, 1979) are examples of controlling the relative magnitude of adhesion and friction in the force system applied by plows to clayey soils. Reduction of adhesion can cause dramatic changes in final soil condition in certain problem soils.

The relative magnitudes at which the different types of failure and rigid body motion are manifest in tillage are highly dependent on initial soil condition, S_i, particularly soil moisture content (Fig. 1). Thus, for a given tool shape, T_s, and manner of movement, T_m, the final soil condition, S_f, is highly dependent on S_i.

Soil-Machine Behavior

The force system imposed on the soil by a tillage tool is more complex than the force system imposed on a soil sample in a soil strength test, such as a triaxial test. This complexity has been a major deterrent to the development of an adequate description of soil behavior. A force boundary condition exists for the soil in a triaxial test, whereas a geometric boundary condition exists for the soil in tillage. The force system, F, for a force boundary condition (triaxial test) is independent of soil behavior, but the geometry of the deformed sample is dependent on soil behavior. The force system, F, that is associated with some geometric boundary condition (such as tillage tool, cone penetrometer, or shear vane), depends on soil behavior and cannot be defined without consideration of at least one or more soil behavioral properties. This fact is considered conceptually in Eq. [3] and [4] (Gill and Vanden Berg, 1967) since tool shape and manner of movement are independent factors. Thus, a tillage tool operated in the same manner in a soil with two initial soil conditions, S_{i1} and S_{i2} (e.g., two moisture states) will apply two different force systems, F_1 and F_2, to the soil. The force system created by a tillage tool is of interest for various reasons.

An engineer is faced with designing tillage tools for use in a wide range of initial soil conditions. Thus, the reliability and durability of the machine's framework and soil working parts are major concerns. Other design concerns include energy requirements, maintenance, and adjustment.

The engineering mechanics and criteria for structural design are well developed. Structural design of a machine (framework and parts) requires knowledge of the force system imposed on the machine. Thus, it was natural for research engineers to pursue the development of predictive equations for the force systems generated by tillage tools to aid structural design.

Engineers must also be concerned with the functional performance of the tillage tool. But the functional performance of a tillage tool—change in soil condition—is not well defined in quantifiable terms. Thus, as emphasized by Spoor (1975), no clearly defined goal has been established for developing equations that will predict soil condition. Consequently, most past research by engineers has been directed toward developing technology for predicting the force system created by a tillage tool (Eq. [3]); and less research has been directed towards relating soil condition and tillage (Eq. [4]). Qualitative and quantitative descriptions of a tillage tool's functional performance are difficult because the final soil condition, S_f, should be adequately defined and quantified based on intended use.

Few theoretical and analytical techniques have been developed for predicting force systems (Eq. [3]) except for tillage tools with simple shapes. Soehne (1956) analyzed the action of an inclined plane tool. He assumed that four soil behavior equations adequately described the tillage action: soil-metal friction, shear failure, acceleration force for each block of soil, and cutting resistance. He reasoned that the soil blocks were formed by shearing, as described by Nichols and Reed (1934). Assuming negligible cutting resistance, Soehne developed an equation that adequately predicted experimental results in a sandy soil, but underpredicted experimental values in loam soil by about 18%.

Rowe and Barnes (1961) modified Soehne's equation to include the influence of adhesion on the soil-metal sliding surface. Their equation was

$$W = G/Z + (CA_1 + B)/[Z(\sin \beta + \nu \cos \beta)]$$

$$+ C_\alpha A_o/[Z(\sin \delta + \mu \cos \delta)], \qquad [7]$$

where
$$Z = [(\cos \delta - \mu \sin\delta)/(\sin \delta + \mu \cos \delta)$$

and
$$+ (\cos \beta + \nu \sin \beta)/(\sin \beta + \nu \cos \beta)] \qquad [8]$$

W = draft (neglecting cutting)
G = weight of soil segment on the tool
C = cohesion of soil
A_1 = area of forward shear failure surface
B = acceleration force of the soil
C_α = soil-metal adhesion
A_0 = area of inclined tool
β = angle of forward shear failure surface
ν = coefficient of internal soil friction
δ = lift angle of the tool
μ = coefficient of soil-metal friction.

Like Soehne's equation, the Rowe and Barnes equation (Eq. [7]) predicted experimental results in sandy soils adequately, but underestimated the experimental results in a silty clay loam by approximately 15%.

Gill and Vanden Berg (1967) and Spoor (1975) presented comprehensive reviews of research in which classical soil mechanics theories, such as the Rankine theory and the Coulomb theory, were used to predict force systems on simple shapes. In most of the studies they reviewed, predictions were adequate for sands and for some clays but were in error for loams, which comprise most agricultural soils. So, at present, no adequate mechanics has been developed for quantitatively describing agricultural soil-machine force systems.

The principles of similitude as utilized in such areas as aerodynamics, thermodynamics, and structural mechanics have been applied in soil-machine research. A knowledge of the behaviors in the physical system is required, but knowledge of the mathematical interrelation of the behaviors is not required. This semiempirical approach, based on the physical modeling of soil-machine systems, has been the subject of considerable study (ASAE, 1977). Some progress has been made and technology has been developed to aid in designing and developing soil-machine systems (Sommer and Wismer, 1979), but the predictions are often inadequate due to insufficient knowledge of soil behavior. Therefore, dimensionless numbers that quantify soil failure and flow (similar to the Reynolds and Mach numbers in aerodynamics) have not been developed and adopted for soil dynamics.

Final Soil Condition

Agricultural soils act as media in which water, air, nutrients, and energy are transmitted to seed and plants; thus, soil behaviors that describe the storage and transmission of these entities are of prime importance. Remember, as Spoor (1975) noted, that plants do not respond to the tillage tool directly, but, rather, to the soil environment created.

Because plant roots provide the contact with the soil that is necessary for the transmission of water, air, nutrients, and energy to a plant, a soil environment and profile conducive to root growth and proliferation are desirable to maximize plant production. Root growth can cause soil deformation in a region around the root tip. As the root grows it must create a force system sufficient to penetrate this region of soil. So, soil strength (a behavioral property) can influence root growth. Thus, transmission and soil strength properties defining soil behavior associated with root growth are important measures of the final condition of soil intended for crop production.

Water, air, and energy enter and exit the soil at the soil-atmosphere interface. This transfer is influenced by soil-surface characteristics. Transfer at the soil-atmosphere interface indirectly depends on the surface geometry. Surface roughness (a state property in contrast to a behavioral property) is influenced by tillage, and has been studied, as reviewed by Soane (1975).

Typical studies of soil surface roughness induced by tillage tools are those by Luttrell et al. (1964), Allmaras et al. (1966), and Currence and Lovely (1970). Currence and Lovely (1970) investigated ways of quantifying soil-surface roughness. They concluded that the method of quantifying surface roughness depends on intended use. Typical values of random roughness, as defined by Allmaras et al. (1966), ranged, in their studies, from about 0.5 cm before tillage to about 3 cm after moldboard plowing. Allmaras et al. (1977) reported that random roughness on the plowed surfaces of a clay loam soil was reduced as much as 40% by rainfall. Thus, both tillage and climatic forces influence surface characteristics.

The directional characteristics (such as ridges) of a tilled surface, which may be important for infiltration and surface runoff predictions, are not quantified by random roughness. Cruse et al. (1980) used random roughness to predict the heat energy flow and balance at the soil surface.

Soil structure within the tilled layer has been studied by several investigators, including Allmaras et al. (1966) and Luttrell et al. (1964). Typical measures of soil structure have been mean clod or aggregate size (mean weight diameter or geometric mean diameter) and changes in both intra-aggregate and interaggregate porosity. Aggregate size and porosity are state properties. Interaggregate porosity tends to reflect tillage effects, whereas intra-aggregate porosity tends to reflect long-term management and soil properties (Larson and Allmaras, 1971). Allmaras et al. (1977) summarized porosity-component data for various tillage tools and combinations of tillage tools used on two soils. They reported that the interaggregate porosity ranged from about 0.1 to 4.3, depending on tillage tool combination and soil. They used state properties (porosity, water content, texture, and organic matter) to estimate the behaviorial properties, thermal diffusivity, heat capacity, and thermal conductivity.

Porosity distribution at a given depth and its variation with depth influence root growth and development (Allmaras et al., 1973). Few results have been reported on this aspect of soil structure. Methods of quantifying porosity distribution are labor intensive and time consuming. Eriksson et al. (1974) and Ojeniyi and Dexter (1979b) have reported results of porosity distribution within the tilled layer.

Ojeniyi and Dexter (1979a) developed a sensitive method of quantifying the internal structure of tilled soil. They found that cropping history and tillage management practices had a great effect on the size of aggregates and voids. When pasture was included in the crop rotation, the structure produced was finer than that for fallow or continuous cereals. They found that a chisel plow produced the maximum number of small aggregates and the minimum number of large voids at a moisture content of about 90 % of the plastic limit. This optimum moisture content was also reported by other investigators (Gill, 1967; Allmaras et al., 1969).

Structural changes induced by multiple-pass tillage were investigated by Ojeniyi and Dexter (1979b). They found that multiple passes of tillage tools had two main effects on soil macro-structure; the multiple passes reduced the aggregate size and sorted the sizes so that the smaller ones tended to migrate toward the bottom of the tilled layer. The second implement pass produced a greater variation of porosity with depth at a moisture of about 130 % of the plastic limit than was produced at a moisture of about 65 % of the plastic limit.

The work of Ojeniyi and Dexter (1979a, 1979b) points out the importance of the management and timeliness of tillage. Soil moisture has a great influence on soil behavior (Fig. 1) that influences the soil condition produced by a tillage tool (Baver et al., 1972).

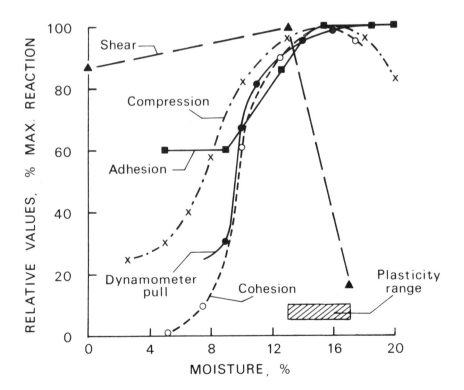

Fig. 1. Relation of dynamic factors involved in tillage to soil moisture with special reference to the plasticity range. (The maximum value for each of these factors was taken as 100.) (Baver et al., 1972).

The research cited in this section illustrates that most often state properties have been used instead of behavioral properties to describe tillage and the soil condition resulting from tillage. These descriptions are adequate if the state properties (e.g., aggregate size, porosity, and moisture content) uniquely define behavioral properties (e.g., soil strength, hydraulic conductivity, and thermal diffusivity). Further, the force boundary conditions and/or geometric boundary conditions have most often been qualitatively defined (e.g., a moldboard plow operating 20 cm deep at 6 km/hour). This qualitative definition of tillage is analogous to a state property description. However, this failure to define quantitatively is not surprising; it merely reflects the state of knowledge of soil dynamics and the soil condition required for agronomic crop production.

Research is underway to develop computer models that will predict soil condition based on the current state of knowledge in soil dynamics. Bowen (1975) is developing a computer model for predicting pressure distribution, porosity, and bulk density in soil under tractor-implement traffic (compaction behavior of soil). He is using behavior equations suggested by Soehne (1958) for describing soil reaction to point and uniform circular loads (force boundary conditions). Research led by Larson and Linden (1980, personal communication. St. Paul, Minn.) has the objective of simulating, with a computer model, the soil condition produced by tillage and climate and the dynamic state of soil moisture and temperature. Computer modeling can indicate the adequacy or deficiencies in our state of knowledge of soil dynamics.

The importance of defining the soil condition needed for optimal crop production and of developing a soil dynamics for prescribing the soil manipulation that will produce the desired soil condition should stimulate us to develop new methods and technologies.

SOIL DYNAMICS AND TILLAGE

To further explore the relation of soil dynamics to tillage, we borrow from a concept developed by the tillage engineering research group at Auburn as an overview of tillage in a broad sense. As they examined tillage systems, they realized that many of them are general and are combinations of operations applied in a broadcast manner without due regard to initial soil condition, S_i, or final soil condition, S_f. They perceived that future tillage systems must be prescribed for specific crops, soil types, soil conditions, and environmental conditions on a narrower geographic scale than is now practiced. Tillage systems will be prescribed just as livestock feed rations are prescribed (Custom Prescribed Tillage, CPT). The CPT concept was described in detail (Johnson, C. E., A. C. Bailey, J. G. Hendrick, C. A. Reaves, and R. L. Schafer. 1980. Custom prescribed tillage. Unpublished report, National Tillage Machinery Laboratory and Auburn Univ., Auburn, Ala.) and only an overview will be presented here.

Custom Prescribed Tillage is presented diagrammatically in Fig. 2 as a systems approach to understanding the role of tillage in agricultural production systems and the entities that are needed to implement the prescribed tillage concept. The CPT would cover the broad spectrum of till-

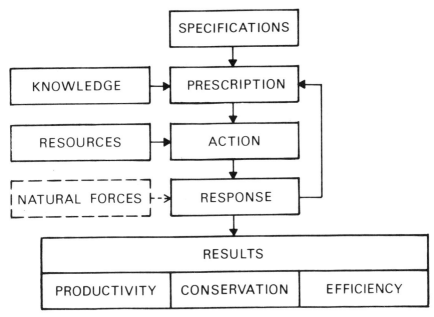

Fig. 2. Custom Prescribed Tillage (CPT).

age from the agricultural production systems that require extensive mechanical soil manipulation to those that require little or no mechanical soil manipulation.

SPECIFICATIONS for a crop form the interface between agronomic needs of the production system and the tillage system. SPECIFICATIONS must be quantified with respect to the seedbed, root zone, traffic lane, water conservation, soil conservation, amendments, pest control, and timeliness. These are the conditions or the behaviors that must be created or maintained by tillage management to implement the production system. SPECIFICATIONS may be a function of time or they may be dynamic for other reasons. Other agronomic aspects of the production system, such as weeds, diseases, and insects may require a dynamic change in SPECIFICATIONS.

KNOWLEDGE is the body of information about soil dynamics, machines, economics, and climate that is coupled with a knowledge of SPECIFICATIONS and RESOURCES to develop a PRESCRIPTION for a tillage system and its management. Thus, soil dynamics must be developed to implement CPT.

Based on SPECIFICATIONS, KNOWLEDGE, and RESOURCES (machines, energy, land, water, climate, time, capital, and labor), a tillage PRESCRIPTION is developed. PRESCRIPTION is a description of the components of the tillage system and how they are managed; e.g., what machines are to be used, how they are to be combined, how they are to be used, the sequence in which they are to be used, and when they are to be used. ACTION is the implementation of the PRESCRIPTION to produce a response. NATURAL FORCES are the dynamic forces of nature, such as weather and biological activity, that act on the production system and influence the RESPONSE. RESPONSE has a feedback of PRESCRIPTION for evaluation of its adequacy and for appropriate adjustment, if needed.

RESULTS are the end product of CPT. RESULTS could be evaluated in terms of meeting the SPECIFICATIONS; however, it may be impossible to meet SPECIFICATIONS exactly because of limits in available KNOWLEDGE and RESOURCES. Traditionally, RESULTS have been evaluated in terms of PRODUCTIVITY, CONSERVATION, and EFFICIENCY, the board of design criteria for agricultural production systems.

Custom Prescribed Tillage is a dynamic control system. A dynamic control system responds to forces imposed on it. The major forcing functions in CPT are SPECIFICATIONS and NATURAL FORCES. To a lesser degree RESOURCES is also a forcing function. Depending on the dynamics of the forcing functions, PRESCRIPTION and ACTION might be quite dynamic. The goal of CPT would be to optimize RESULTS.

Custom Prescribed Tillage is practiced to some extent, very crudely, in current agricultural production systems. It is more fully developed in areas where the climate is predictable and the soil is uniform than in areas where both climate and soil are highly variable. However, some elements in CPT, particularly soil dynamics, are not sufficiently defined for implementing this approach in production agriculture.

Whether one views tillage in the CPT perspective or in some other perspective, the goals described by CPT have long been sought. A good soil dynamics is essential to CPT, to similar concepts, and to modeling because the reaction of soil to force systems is essential to prescribing actions that will produce the conditions specified.

SUMMARY

Hopefully, this discussion has been sufficiently descriptive of the state of knowledge available to answer the four questions that authors were asked to address. In summary, some of the soil and soil-machine behaviors that are manifest in tillage have been identified. Quantitative descriptions that can be used in predictive equations have been developed for some behaviors, but not for others. Quantitative descriptions of the complex interaction of these behaviors in tillage and of the soil condition have not been developed. A comprehensive soil dynamics is needed as the basis for designing tillage systems that will produce a soil condition that has been prescribed for crop production.

The efficiency and effectiveness of present-day tillage systems can not be disputed. These systems have been developed based on astute observations and past research. The history of technological development has indicated that science often follows in the footsteps of invention. Someone determines how to make something work (invention), and then others determine why it worked (science) and how to make it work better (development). However, current tillage systems have been developed based more on qualitative descriptions of soil and machine behavior than on quantitative descriptions. Perhaps further improvements (inventions), dictated by necessity, may be made without a soil dynamics that quantitatively describes tillage. However, science and development may lead to greater improvements. We are not saying that new tillage concepts should not be explored until a soil dynamics is developed. Certainly, we have used the best knowledge available to enhance the effectiveness and efficiency of tillage systems, and we must continue to do so. However, one has to wonder what could be done if a soil dynamics was available for the design of tillage systems. Certainly, we have excellent examples of the impact of engineering sciences on the design of other physical systems.

LITERATURE CITED

1. Allmaras, R. R., A. L. Black, and R. W. Rickman. 1973. Tillage, soil environment, and root growth. p. 62–86. *In* Proc. National Conserv. Tillage Conf., Des Moines, Iowa. March 1973. Soil Conserv. Soc. of Am., Ankeny, Iowa.

2. ――――, R. E. Burwell, and R. F. Holt. 1969. Plowlayer porosity and surface roughness from tillage as affected by initial porosity and soil moisture at tillage time. Soil Sci. Soc. Am. Proc. 31:550–556.

3. ――――, ――――, W. E. Larson, and R. F. Holt. 1966. Total porosity and random roughness of the interrow zone as influenced by tillage. USDA-ARS Conserv. Res. Rep. no. 7.

4. ――――, E. A. Hallauer, W. W. Nelson, and S. D. Evans. 1977. Surface energy balance and soil thermal property modifications by tillage-induced soil structure. Minnesota Agric. Exp. Stn. Tech. Bull. 306-1977.

5. American Society of Agricultural Engineers. 1977. Similitude of soil-machine systems. Publ. 3-77. Am. Soc. Agric. Eng., St. Joseph, Mich.

6. Bailey, A. C., and J. A. Weber. 1965. Comparison of methods of measuring shear strength using artificial soil. Trans. Am. Soc. Agric. Eng. 8:153–156, 160.

7. Baver, L. D., W. H. Gardner, and W. R. Gardner. 1972. Soil physics. John Wiley and Sons, Inc., New York.

8. Bowen, H. D. 1975. Simulation of soil compaction under tractor-implement traffic. Paper no. 75-1509. Am. Soc. Agric. Eng., St. Joseph, Mich.

9. Cooper, A. W., and W. R. Gill. 1966. Characterization of soil related to compaction. Grundforbattring AGR 19:77–80 NRI, Uppsala, Sweden.

10. ――――, and W. F. McCreery. 1961. Plastic surfaces for tillage tools. Paper no. 61-649. Am. Soc. Agric. Eng., St. Joseph, Mich.

11. Cruse, R. M., D. R. Linden, J. K. Radke, W. E. Larson, and K. Larntz. 1980. A model to predict tillage effects on soil temperature. Soil Sci. Soc. Am. J. 44:378–383.

12. Currence, H. D., and W. G. Lovely. 1970. The analysis of soil surface roughness. Trans. Am. Soc. Agric. Eng. 13:710–714.

13. Dunlap, W. H., G. E. Vanden Berg, and J. G. Hendrick. 1966. A comparison of soil shear values obtained with devices of different geometrical shapes. Trans. Am. Soc. Agric. Eng. 9:896–900.

14. ――――, and J. A. Weber. 1971. Compaction of an unsaturated soil under a general state of stress. Trans. Am. Soc. Agric. Eng. 14:601–607, 611.

15. Eriksson, J., I. Hakansson, and B. Danfors (English translation by J. K. Aase). 1974. The effect of soil compaction on soil structure and crop yields. Swedish Inst. Agric. Eng. Bull. 354, Uppsala, Sweden.

16. Gill, W. R. 1967. Soil-implement relations. p. 32–36, 43. *In* Proc. Tillage for Greater Crop Production Conf. Am. Soc. Agric. Eng., St. Joseph, Mich. December 1967.

17. ――――, and G. E. Vanden Berg. 1967. Soil dynamics in tillage and traction. USDA Agric. Handb. no. 316. U.S. Government Printing Office, Washington, DC.

18. Hendrick, J. G., and G. E. Vanden Berg. 1961. Strength and energy relations of a dynamically loaded clay soil. Trans. Am. Soc. Agric. Eng. 4:31–32, 36.

19. Higdon, A., and W. B. Stiles. 1957. Engineering mechanics. Prentice-Hall, Inc., Englewood Cliffs, N.J.

20. Kumar, L., and J. A. Weber. 1974. Compaction of unsaturated soil by different stress paths. Trans. Am. Soc. Agric. Eng. 17:1064–1069, 1072.

21. Larson, W. E., and R. R. Allmaras. 1971. Management factors and natural forces related to compaction. p. 367–427. In K. K. Barnes, W. M. Carleton, H. M. Taylor, R. I. Throckmorton, and G. E. Vanden Burg (ed.) Compaction of agricultural soils. Am. Soc. Agric. Eng., St. Joseph, Mich.

22. Luttrell, D. H., C. W. Bockhop, and W. G. Lovely. 1964. The effect of tillage operations on soil physical conditions. Paper no. 64-103. Am. Soc. Agric. Eng., St. Joseph, Mich.

23. Nichols, M. L. 1931. The dynamic properties of soil. II. Soil and metal friction. Agric. Eng. 12:321–324.

24. ————, and I. F. Reed. 1934. Soil dynamics: VI. Physical reactions of soils to moldboard surfaces. Agric. Eng. 15:187–190.

25. Ojeniyi, S. O., and A. R. Dexter. 1979a. Soil factors affecting the macro-structures produced by tillage. Trans. Am. Soc. Agric. Eng. 22:339–343.

26. ————, and ————. 1979b. Soil structural changes during multiple pass tillage. Trans. Am. Soc. Agric. Eng. 22:1068–1072.

27. Rowe, R. J., and K. K. Barnes. 1961. Influence of speed on elements of draft on a tillage tool. Trans. Am. Soc. Agric. Eng. 4:55–57.

28. Schafer, R. L., W. R. Gill, and C. A. Reaves. 1975. Lubrication of soil-metal interfaces. Trans. Am. Soc. Agric. Eng. 18:848–851.

29. ————, ————, and ————. 1979. Experiences with lubricated plows. Trans. Am. Soc. Agric. Eng. 22:7–12.

30. Soane, B. D. 1975. Studies on some soil physical properties in relation to cultivations and traffic. p. 160–182. In Soil physical conditions and crop production. Ministry of Agriculture, Fisheries, and Food Tech. Bull. 29. London.

31. Soehne, W. H. 1956. Einige Grundlagen fur eine landtechnische Bodenmechanik. (In German) Grundlagen der Landtechnik 7:11–27.

32. ————. 1958. Fundamentals of pressure distribution and compaction under tractor tires. Agric. Eng. 39:276–281, 290.

33. Sommer, M. S., and R. D. Wismer. 1979. Application of soil dynamics technology in the design of agricultural and industrial equipment. Paper no. 79-1544. Am. Soc. Agric. Eng., St. Joseph, Mich.

34. Spoor, G. 1975. Fundamental aspects of cultivations. p. 128–144. In Soil physical conditions and crop production. Ministry of Agriculture, Fisheries, and Food Tech. Bull. 29. London.

35. Steel, R. G. D., and J. H. Torrie. 1960. Principles and procedures of statistics. McGraw-Hill Book Co., New York.

36. Vanden Berg, G. E., and C. A. Reaves. 1966. Characterization of soil properties for tillage tool performances. Grundforbattring AGR 19:49–58 NRI, Uppsala, Sweden.

37. Yong, R. N., and B. P. Warkentin. 1966. Introduction to soil behavior. MacMillan Co., New York.

Chapter 3

Tillage Effects on the Hydraulic Properties of Soil: A Review[1]

A. KLUTE[2]

ABSTRACT

The water retention, hydraulic conductivity, and diffusivity of soils as functions of water content and/or suction are the hydraulic properties of soils, and play an important central role in determining the movement and storage of water in soil. The general purpose of tillage is to create a soil environment favorable to the desired plant growth. Soil water relations are an important aspect of the soil environment of the plant. A review of the literature was made for data on the effects of tillage and other soil structural modifications on the hydraulic properties. The reported effects are somewhat scattered and often apparently contradictory. Tillage operations modify the bulk density (i.e., porosity) and pore size distribution of the soil. These properties are highly determining factors for the hydraulic properties. The data found in the literature are analyzed in relation to various measures of the pore space geometry such as porosity and pore size distribution.

INTRODUCTION

The ability of soils to retain and transmit water is measured by the hydraulic properties of the soil. These properties are determined by the geometry of the pore space. The latter is modified in various ways by till-

---

[1] Presented at a symposium, "Predicting Tillage Effects on Soil Physical Properties and Processes" at the Am. Soc. of Agron. meeting in Detroit, Michigan, 3 Dec. 1980.

[2] Soil scientist, USDA-ARS and professor of Soils, Colorado State University, Fort Collins, CO.

age operations. The purpose of this paper is to review and discuss tillage effects on hydraulic properties, evaluate the state of our knowledge of such effects and, if possible, identify gaps in that knowledge and some directions that might be taken in research on tillage effects on hydraulic properties.

Hydraulic Properties of Soil

The hydraulic properties are those functions that characterize the water retention and transmission properties of a soil. In the usual Darcy-based theory of water flow in unsaturated soil (e.g., see Klute, 1973) there are two basic hydraulic functions, the hydraulic conductivity K and the water capacity, C. The hydraulic conductivity may be considered a function of water content [$K(\theta)$], or the capillary pressure head of the soil water [$K(h)$]. The water capacity is the rate of change of water content with the capillary pressure head, $d\theta/dh$, and is derivable from the water retention function $\theta(h)$. When the soil is uniform with respect to the water content-pressure head relationship, and when the flow is nonhysteretic, the concept of soil water diffusivity, $D = K/C$, may be used. In most cases, the diffusivity is treated as a function of the water content. As usually measured, the hydraulic conductivity and soil water diffusivity will include a contribution due to vapor phase transport of water, especially at lower water contents (Philip, 1957; Rose, 1963). In this paper, the term *hydraulic properties* refers to all or a part of the following set of functions: $K(\theta)$, $K(h)$, $\theta(h)$, $C(h)$, $C(\theta)$, and $D(\theta)$. In most instances, attention will be concentrated upon $\theta(h)$ and $K(\theta)$.

The significance of the hydraulic properties lies in their use in a quantitative analysis of water transport in the soil profile. Thus, they offer the possibility of quantitatively and rationally assessing the influence of tillage-induced changes in the hydraulic properties upon the water regime in the soil profile.

Problems of Definition of Hydraulic Properties

The hydraulic properties are assumed to be definable in the sense of the macroscopic continuum approach to the description of flow in porous media. In that approach the actual porous medium, with the fluids contained in its pores, is replaced by a fictitious continuum. At each mathematical point of this continuum, macroscopic parameters such as porosity, conductivity, water capacity, water content, and pressure head are assumed to be definable. The values of the parameters are obtained by averaging the appropriate local microscopic parameters of the medium over a small porous medium volume, i.e., a representative elementary volume (REV). In this way, the local inherent inhomogeneity, resulting from the pore or grain size distribution, is replaced by a continuum description of the medium.

In a soil with inhomogeneity due only to the primary grain (and pore) structure the inhomogeneity is characterized by a length that is of the order of the mean grain or pore size. For such media the characteristic length of the REV will be somewhat larger than the mean pore size. The pore size distribution is in this case generally uni-modal. For soils with grain sizes of agricultural interest (< 2 mm), the REV will usually be sufficiently small that one is not distressed by considering it as a differential volume element of the medium located at a point in the medium in the macroscopic sense. Measurement devices for soil water potential such as piezometers, tensiometers, and methods of determining the water content produce data that seem to offer little if any difficulty in associating the results with a given point in the medium.

In many soils, a higher order of inhomogeneity is found which is due to the presence of fractures, fissures, channels, root holes, aggregates, clods, etc. Tillage commonly produces such inhomogeneities. For such media the length characterizing the inhomogeneity is of the order of the dimensions of the clods or aggregates, or of the separation of the fractures or channels. The REV for these soils will necessarily be larger than that required for soils of a single grain structure. The REV may easily be so large that one is reluctant to treat such a volume element as differential. The large REV creates problems in defining the macroscopic parameters of the fictitious continuum that are preferred in treating flow processes in the soil. Instruments used for measurements of variables such soil water content and potential may not yield appropriate macroscopic average values that can be associated with the large REV required for these bi and polymodal soils. Samples of large physical size may be needed to obtain appropriate averages. In such media the usual Darcy-based flow theory is not directly applicable, and at least requires significant modification. The concepts and consequences of flow in macropores and fractures have recently been discussed by Bear and Braester (1972), and Thomas and Phillips (1979) among others. These concepts must not be forgotten in any consideration of the effects of tillage on the hydraulic properties of soil.

Hydraulic Properties Measured in Tillage Studies

Relatively few studies of tillage seem to have been conducted in which measurements of hydraulic properties were made. The complexities and difficulties of adequate sampling, and the techniques for determining the hydraulic functions have tended to discourage researchers from making the measurements. Simpler, more easily determined parameters, such as bulk density, have usually been used to assess the effects of tillage practices. Some selected reports of tillage studies in which hydraulic properties were measured are reviewed below.

The effect of tillage practices vs. no-till on the hydraulic properties of a grey-brown podzolic soil (Eutroboralk) derived from loess, has been extensively studied by Ehlers and his associates at Goettingen (Ehlers, 1976, 1977; Ehlers and van der Ploeg, 1976a, 1976b). Conductivity and water retention functions were measured in situ. Soil water diffusivity was determined on soil core samples by a rapid laboratory evaporation method developed by Arya et al. (1975). Water retention data were also obtained on core samples. The hydraulic conductivity was calculated from the soil water diffusivity and the water capacity. The latter was obtained from the water retention data. The soil water diffusivity in the 10 to 20 cm layer of the untilled soil was higher than that in the corresponding layers of the tilled soil at water contents greater than about 0.34 cm^3/cm^3. At lower water contents the soil water diffusivity was higher in the 10 to 20 cm layer of the tilled soil. These results are shown in Fig. 1. Hydraulic conductivity-soil moisture tension data are shown in Fig. 2. For a given soil layer the conductivity of the untilled soil at low suction was higher than that of the tilled soil. This result is to be expected on the basis of the destruction of macroporosity in the tilled soil. The conductivity of the 20 to 30 cm layer of the tilled soil, which is at the bottom of the plow layer, was much lower than that of the corresponding layer of the untilled soil and reflects the compaction of the soil at this depth induced by plowing.

Another example of a tillage study in which hydraulic properties of the soil were measured is that of Allmaras et al. (1977). Soil hydraulic properties were measured in situ by an instantaneous profile method to evaluate the effect of chiseling on the water regime in a Walla Walla silt loam (coarse-silty, mixed, mesic Typic Haploxeroll). The water retention in the -50 to -300 mb range of the chiseled soil was slightly reduced relative to that of the untilled soil. At soil water pressure heads greater than -50 mb the water retention function for the two treatments was unchanged. These results were found for the 10, 20, 30, and 40 cm depths. Hydraulic conductivity-water content data for the 10, 20, and 30 cm depths showed that chiseling increased the conductivity at lower water contents (less than about 0.36 cm^3/cm^3). At higher water contents there was a tendency for the conductivity of the untilled soil to be greater than that of the chiseled soil for the same three depths. At depths of 40 cm and

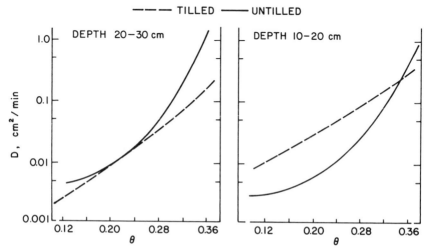

Fig. 1. Soil water diffusivity—water content functions for two layers of tilled and untilled soil. Adapted from Ehlers (1976).

greater the $K(\theta)$ functions were unaffected by chiseling as would be expected since the chiseling depth was approximately 40 cm. In this study, chiseling produced a soil with a greater degree of aggregation, which may be reflected in these changes in $K(\theta)$. A theoretical analysis (Farrell, 1972) of the conductivity of aggregated vs. compacted unaggregated soil showed that the conductivity of the aggregated medium at a given water content should be higher which is qualitatively in agreement with the results obtained in the chiseling study.

England (1971) gives a summary of results for the water retention curves of the surface soil of Mollisols and Alfisols under pasture and row-crop cultivation. The data show that below 1/3 bar suction cultivated Mollisols retained 40% more water and cultivated Alfisols retained 25% more water than corresponding pastured soils. At suctions above 1/3 bar, cultivated soils of both orders held less water than the pastured soils. The soils were silt loams. Cultivation of both soil orders resulted in an increase in the proportion of large pores. The water retained between 1/3 and 15 bars was less for cultivated soils than for pastured soils.

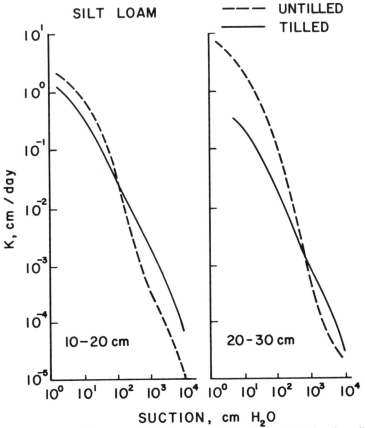

Fig. 2. Hydraulic conductivity—suction functions for two layers of tilled and untilled soil. Adapted from Ehlers (1976).

Bouma et al. (1975) have studied the hydraulic properties and soil morphology of the principal horizons of paired virgin and cultivated pedons of two soils, Tama silt loam and Oshkosh clay. Cultivation had been imposed for approximately a century. Water retention was measured on core samples. Saturated hydraulic conductivities were measured in situ with the Bouwer double tube method (Bouwer, 1961). Hydraulic conductivity-suction relations were calculated from the water retention data by the method of Green and Corey (1971). The in situ measured saturated conductivity data was used to develop a matching factor. The saturated conductivity of the horizons in cultivated pedons were lower than those of the virgin horizons due to the decrease in number of large pores. Except for the A_p and A_1 horizons of the Tama silt loam, the conductivity of the horizons of the cultivated pedons was higher than that of the virgin pedons above certain moisture tensions, i.e., the conductivity-tension curves of corresponding horizons of cultivated and virgin pedons crossed at some value of tension. Figure 3 shows selected examples of these results. These differences reflect changes in pore size distribution induced by cultivation that increased the relative volume of fine pores.

Tillage and Pore Space Geometry

Tillage generally tends to decrease the bulk density and increase the total porosity of the surface soil. At the same time, the soil just below the plowed or tilled layer may be increased in bulk density by the stresses applied to that layer by tillage machinery. The pore space geometry produced in the surface soil is usually very unstable and changes of the pore space geometry with time are common.

Tillage usually produces changes in the pore size distribution of the soil. A detailed description of the pore space geometry is impossible, and

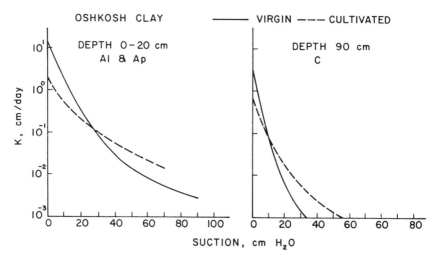

Fig. 3. Hydraulic conductivity—suction functions for two horizons of virgin and cultivated Oshkosh clay. From Bouma et al. (1975).

only a statistical description, such as a pore size distribution function (PSD), is possible. The PSD function represents the volume (or volume function) of pores of a given size as a function of the size. Because of the irregular geometry of the pore space, no single dimension of a pore can unambiguously be identified as its size. The pore size must be defined by a method of measurement. Often capillary concepts are applied to estimate a pore size distribution from the displacement of the wetting fluid (the soil water) from the soil by a non-wetting fluid (air). This leads to the concept that a water retention curve, $\theta(h)$, can be used to develop a PSD function. The approach is most applicable in the wetter range, where surface tension-radius of curvature capillary phenomena are dominant in the retention of the soil water. At lower water contents where short-range absorptive forces between the solid and liquid phases dominate the retention of the water, the approach is less applicable.

The PSD function may display one or more mixima or peaks. A uni-modal PSD function is characteristic of a large number of soils, especially sandy soils with single grain structure. Bi-modal (and perhaps poly-modal) pore size distributions are found in soils with well developed structure. These structural units are commonly clods, and aggregates in the surface soil. In the subsoil there may be peds with fracture zones between them that contribute to a bi-modality of the pore space. Worm holes and root channels in both surface and subsoil also contribute to the bi-modality of the PSD.

In soils with a uni-modal PSD the tillage process may produce changes in the mean pore size and the dispersion about the mean. Compaction and increase of bulk density requires a decrease on the total porosity and decreases the mean pore size. The fraction of the porosity that is made up of larger pores is reduced. These effects are displayed schematically in Fig. 4.

A tillage process may also change the PSD of the soil from unimodal to bi-modal. Such is often the case in tillage of the surface soil which produces clods and aggregates. Conversely, it is also possible that the tillage process may destroy macropores that were present in the untilled soil and produce a more uni-modal PSD (e.g., see Ehlers, 1975; Wilkinson and Aina, 1976).

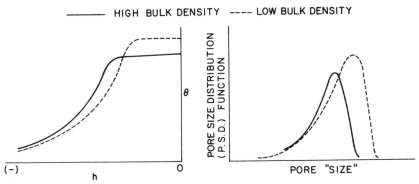

Fig. 4. Schematic of changes in the water retention function and associated pore size distribution function produced by increased bulk density.

Hydraulic Properties and Pore Space Geometry

The changes in the geometry of the pore space produced by tillage, and reflected in the total porosity and PSD, will have important effects on the hydraulic functions. Reduction of total porosity obviously will reduce the volumetric water content of the soil at zero suction. The destruction of macropores will reduce the conductivity at saturation since such pores when waterfilled can contribute strongly to the water flow through soil in response to an imposed hydraulic gradient. Since such macropores will drain at relatively low suctions there will be large decreases in conductivity as the soil water suction increases from zero.

Calculations of the hydraulic conductivity from moisture retention data have often been made using one or another of the procedures in the literature (Childs and Collis-George, 1950; Marshall, 1958; Millington and Quirk, 1961; Laliberte et al., 1968; Mualem, 1978). A fundamental assumption in these methods is that a pore size distribution can be obtained from the water retention data. Most of these theories seem to have been based on media with a uni-modal pore size distribution, and few estimates of conductivity for bi-modal PSD media have been made. There are difficulties with the methods of calculation, which have been identified by various authors (e.g., see Mualem, 1978 and Denning et al., 1974). The latter authors found that agreement between calculated conductivity functions and measured conductivity functions was good on sandy apedal soils, but only if a matching factor was used. Agreement between calculated and measured functions was not as good on clayey pedal soil horizons in which a few relatively large planar and tubular pores determined K in the wet range. In such media, the greatest fraction of total porosity is in the fine pores within the peds, and these contribute very little to the flow.

Denning et al. (1974) do not recommend the use of calculation procedures for determining the conductivity functions of pedal soils, i.e., those with bimodal PSD functions. The same conclusion would seem to apply to the estimation of hydraulic conductivity of surface tilled soil in which a strongly bimodal PSD has been developed.

The water retention and transmission properties of aggregated soil materials has been investigated by Wittmus and Mazurak (1958), Farrell (1972), Tamboli et al. (1964), and Amemiya (1965). Since tillage can create an aggregated condition of the soil, some insight into tillage effects on soil hydraulic properties may be obtained by considering the results of these studies. From these studies it appears, as would be expected, that a packing of aggregates of a given size range will hold more water at zero suction than a packing of primary particles of the same size range. This is due to the additional porosity within the aggregates. The aggregate packing has a bimodal PSD. The interaggregate pore space will be approximately of the same size as that for primary particles of the same size as the aggregates, and will drain at about the same suction. The drainage of the intra-aggregate porosity is somewhat more questionable and would depend on the degree of hydraulic interconnection of the water held in the internal pore space of the aggregates. In measurements of water retention of soils by suction or pressure cell apparatus, most soil materials (especially the coarser textured soils) display a residual water content which is very slow to be removed. In packings of primary particles the residual water is located at points of contact between grains and is very poorly interconnected. Removal of this water can occur by vapor movement and slow film flow. It seems reasonable to expect that packings of aggregates might tend to hold more residual water than a packing of primary particles of the same size range. Water should be held at contact points of the aggregates, and in addition there would be water present in the internal pore space of the aggregates. If such water is poorly interconnected in the hydraulic sense it would contribute to a higher residual water content. However, it is also possible that the internal porosity of the aggregates could provide an improved pathway for removal of the water at higher suctions and thus reduce the residual water content.

Hydraulic Properties and Bulk Density

Since tillage can produce changes in bulk density, some insight into the effects of tillage on hydraulic properties might be gained by examining the effects of bulk density on those properties.

Water Retention and Bulk Density. The effect of bulk density on the water retention function has been examined by many investigators (e.g., see Laliberte et al., 1966; Yang and de Jong, 1971; Croney and Coleman, 1954; Reeve et al., 1973; Hill and Summer, 1967; Archer and Smith, 1972; Box and Taylor, 1962; Campbell and Gardner, 1971). The studies may be classified into two types: (1) those in which the soil was packed to a given bulk density followed by determination of the water retention curve and (2) those in which a soil at a given water content and bulk density was subjected to a compressive stress and the changes in soil water potential were observed.

It has long been recognized that compaction affects the pore size distribution in the larger pore size range more than in the fine. Croney and Coleman (1954) described the changes in the water retention properties of incompressible soils compacted to different initial dry bulk density. In such soils, increasing the bulk density decreases the total porosity and thus decreases the amount of water held at low suctions, and tends to increase the amount of water held at higher suctions. The hysteresis loop of the $\theta(h)$ function was narrower in the more compacted soil.

One of the more systematic studies of the effects of bulk density on the hydraulic properties of porous media is that of Corey and his associates at Colorado State University. Much of the data obtained in that study is reported in Laliberte et al. (1966). They used the Brooks-Corey retention function:

$$(\theta - \theta_r)/(\theta_s - \theta_r) = (P_b/P_c)^\lambda, \qquad P_c \geq P_b \qquad [1]$$

$$\theta = \theta_s \qquad\qquad\qquad P_c < P_b$$

where θ_s is the total porosity of the soil, P_b is the bubbling pressure, P_c is the capillary pressure, θ_r is the residual water content, and λ is a pore size distribution index. They found that packing a soil to a higher bulk density increased the bubbling pressure, P_b, and decreased θ_s. Little effect on λ or θ_r was noted. Figure 5 shows a plot of P_c vs. θ for Touchet silt loam that illustrates these effects of bulk density on the retention curve. Yang and de Jong (1971) give the water retention data for two soils, an Oxbow loam, and a Melfort Clay, which show trends similar to those found by Laliberte et al. (1968).

In experiments in which a soil at a given fixed water content was subjected to compressive stress at constant water content, Box and Taylor (1962), and Campbell and Gardner (1971) found that the moisture potential tended to increase, i.e., the soil water suction was decreased as the bulk density was increased.

Bulk Density and Hydraulic Conductivity. Data on the hydraulic conductivity at various bulk densities have been reported by Douglas and McKyes (1978), van Schaik and Laliberte (1969), Laliberte and Brooks (1967), and Laliberte et al. (1966). In terms of the Brooks-Corey conductivity function:

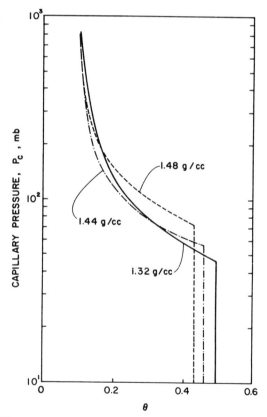

Fig. 5. Capillary pressure—water content functions for Touchet silt loam at three bulk densities. From Laliberte et al. (1966).

$$K = K_s (P_b/P_c)^n \qquad P_c \geq P_b$$

$$K = K_s \qquad\qquad P_c < P_b$$

[2]

it was found that packing a given soil to higher bulk density decreased the saturated conductivity K_s, increased the bubbling pressure P_b and had little effect on n. An example of these results is displayed in Fig. 6, for Touchet silt loam at three bulk densities.

The hydraulic conductivity-water potential function of a clay soil was studied by Douglas and McKyes as a function of the compaction. The hydraulic conductivity increased as the porosity increased at all levels of suction. Van Schaik and Laliberte (1969) present hydraulic conductivity-capillary pressure data for a Chin silty clay loam at two bulk densities which is in qualitative agreement with that obtained by Laliberte et al. (1966).

Bulk Density and Soil Water Diffusivity. In many studies of soil water diffusivity, the soils used have been compacted to a bulk density that was convenient for the experimental conditions at hand. Conse-

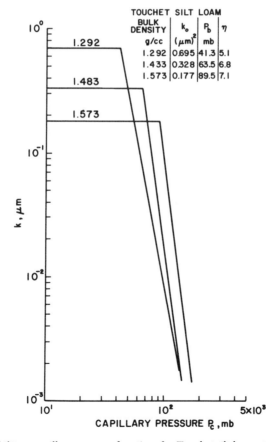

TOUCHET SILT LOAM			
BULK DENSITY	k_0	P_b	η
g/cc	$(\mu m)^2$	mb	
1.292	0.695	41.3	5.1
1.433	0.328	63.5	6.8
1.573	0.177	89.5	7.1

Fig. 6. Permeability—capillary pressure functions for Touchet silt loam at three bulk densities. From Laliberte et al. (1966).

quently, the diffusivity data reported are valid for only a single bulk density. A few studies have examined the effect of bulk density on soil water diffusivity. Gumbs and Warkentin (1972) and Shanda (1977) present data showing that increasing the bulk density decreases the diffusivity at a given water content of a clay soil and a silty clay loam, respectively. Jackson (1963) gives soil water diffusivity data for three soils at several porosities (bulk densities). The diffusivity of Adelanto loam and Pine silty clay was decreased as the bulk density increased. The diffusivity of Pachappa loam was not much affected.

SUMMARY

In summary of the foregoing review of the changes of hydraulic properties and their relation to tillage and various aspects of the pore space geometry, the following changes can be expected in the hydraulic properties:

If the tillage operation produces an increase in the bulk density of a uni-modal PSD soil, (1) the water content at or near zero suction will decrease, (2) the saturated K will be decreased perhaps by a factor of 2 to 5 or even more, (3) the suction at which a continuous air phase is first formed upon drainage will increase, (4) there will be little effect on the residual water content and the pore size index in the Brooks-Corey functions, and (5) the soil water diffusivity will be decreased by a factor of 2 to 10 at a given water content.

If, on the other hand, the tillage operation creates a bimodal PSD with increased total porosity from an essentially uni-modal PSD soil we may expect that (1) the conductivity at zero suction will be very much increased, (2) the water retention at low suctions will be increased, and (3) the imposition of small suctions will cause large reductions in conductivity, and water content as the macropores are drained.

SOME FUTURE DIRECTIONS

The most rational and quantitative analysis of the effect of tillage practices and treatments on the soil water regime would appear to be possible by the use of one or more models of the soil water transport, with associated hydraulic properties reflecting the effect of the tillage. A problem that is beginning to receive more attention is that of appropriate modeling of flow in macroporous, bimodal soils. Such efforts should continue. Questions to be explored include an examination of whether or not modifications to the present Darcy-based flow theory can adequately treat flow in bimodal soils, and the development of new approaches. One approach to modeling flow in macroporous media considers the structural units (peds, aggregates, and clods) to act as sources or sinks for the more mobile water in the macropores. Another approach considers the bimodal medium as two superimposed continua, each with its own hydraulic functions. In this approach the water in the structural units is considered mobile, but characterized by a conductivity that is smaller than that of the macroporous regions at or near zero suction, and larger at higher suctions. In either case, a model is needed to represent the interchange between water in the macropores and that in the structural units.

The problems of definition of the hydraulic function of macroporous media have been identified above. Appropriate methodology that considers the relatively large REV of such media and provides measured hydraulic functions consistent with the particular model to be used should be devised, and applied in the evaluation of the water movement in tilled soils. Such is, of course, the case in any study of soil water movement. The often unstable condition of tilled soil imposes additional difficulties of assessment of soil hydraulic properties. Most of the measurements of hydraulic properties in connection with tillage studies have been made after the soil structure has somewhat stabilized. Consequently, the results cannot fully reflect (if at all) the water retention and transmission properties of the freshly tilled soil. Even though the unstable condition does not prevail for very long, the water transport during this period might be a significant part of the total water regime.

The measurement of hydraulic properties should be conducted on samples that properly represent the pore space geometry of the field soil. This generally precludes the use of disturbed sieved soils that are repacked except possibly for surface soils. Such disturbance modifies and even destroys natural structural units that may be present in the soil in the field. Sample size should be chosen with proper consideration of the size scale of the structural units of the soil in situ. In addition, field variability of the properties being measured must also be considered.

LITERATURE CITED

1. Allmaras, R. R., R. W. Rickman, L. G. Ekin, and B. A. Kimball. 1977. Chiseling influences on soil hydraulic properties. Soil Sci. Soc. Am. J. 41:796–803.

2. Amemiya, M. 1965. The influence of aggregate size on soil moisture content-capillary conductivity relations. Soil Sci. Soc. Am. Proc. 29:744–748.

3. Archer, J. R., and P. D. Smith. 1972. The relation between bulk density, available water capacity and air capacity of soils. J. Soil Sci. 23:475–480.

4. Arya, L. M., D. A. Farrell, and G. R. Blake. 1975. A field study of soil water depletion patterns in presence of growing soybean roots. I. Determination of hydraulic properties of the soil. Soil Sci. Soc. Am. Proc. 39:424–430.

5. Bear, Jacob. 1972. Dynamics of fluids in Porous Media. American Elsevier, N.Y.

6. ————, and Carol Braester. 1972. On the flow of two immiscible fluids in fractured porous media. p. 177–202. In Fundamentals of transport phenomena in porous media. Inter. Assoc. for Hydraulic Research. Elsevier Publishing Co., N.Y.

7. Bouma, J., and J. L. Anderson. 1973. Relationships between soil structural characteristics and hydraulic conductivity. p. 77–106. In R. R. Bruce et al. (ed.) Field soil water regime. S.S.S.A. Spec. Pub. No. 5. Madison, Wis.

8. ————, D. J. van Rooyen, and F. D. Hole. 1975. Estimation of comparative water transmission in two pairs of adjacent virgin and cultivated pedons in Wisconsin. Geoderma 13:73–88.

9. Bouwer, H. 1961. A double tube method for measuring hydraulic conductivity of soil in situ above a water table. Soil Sci. Soc. Am. Proc. 29:334–339.

10. Box, J. E., and I. A. Taylor. 1962. Influence of soil bulk density on matric potential. Soil Sci. Soc. Am. Proc. 26:119–123.

11. Campbell, G. S., and W. H. Gardner. 1971. Psychrometric measurement of soil water potential: Temperature and bulk density effects. Soil Sci. Soc. Am. Proc. 35:8–12.

12. Childs, E. C., and N. Collis-George. 1950. The permeability of porous materials. Proc. R. Soc. Am. 201:392–405.

13. Croney, D., and J. D. Coleman. 1954. Soil structure in relation to soil suction (pF). J. Soil Sci. 5:75–84.

14. Denning, J. L., J. Bouma, O. Falayi, and D. J. van Rooyen. 1974. Calculation of hydraulic conductivities of horizons in some major soils in Wisconsin. Geoderma 11:1–16.

15. Douglas, E., and E. McKyes. 1978. Compaction effects on the hydraulic conductivity of a clay soil. Soil Sci. 125:278–282.

16. Ehlers, W. 1975. Observations on earthworm channels and infiltration on tilled and untilled loess soil. Soil Sci. 119:242–249.

17. ————. 1976. Rapid determination of undisturbed hydraulic conductivity in tilled and untilled loess soil. Soil Sci. Soc. Am. J. 40:837–840.

18. ————. 1977. Measurement and calculation of hydraulic conductivity in horizons of tilled and untilled loess-derived soil, Germany. Geoderma 19:293–306.

19. Ehlers, W., and R. R. van der Ploeg. 1976a. Simulation of infiltration into tilled and untilled field soils derived from loess. p. 157–167. In G. C. Vanoteenkiste (ed.) System simulation in water resources.

20. ————, and ————. 1976. Evaporation, drainage and unsaturated hydraulic conductivity of tilled and untilled fallow soil. Z. Pflanzenernaehr Bodenkd. 3:373–386.

21. England, C. B. 1971. Moisture retention of cultivated and pastured Mollisols and Alfisols. Soil Sci. Soc. Am. Proc. 35:147–149.

22. Farrell, D. A. 1972. The effect of soil crusts on infiltration: The effect of aggregate size and depth of tillage on steady infiltration through crust-topped tilled soils. Meded. Fac. Landbouwwet. Rijksuniv. Gent. 37:1132–1149.

23. Green, R. E., and J. C. Corey. 1971. Calculation of hydraulic conductivity: a further evaluation of some predictive methods. Soil Sci. Soc. Am. Proc. 25:3–8.

24. Gumbs, F. A., and B. P. Warkentin. 1972. The effect of bulk density and initial water content on infiltration in clay soil samples. Soil Sci. Soc. Am. Proc. 36:720–724.

25. Hill, J. N. B., and M. E. Summer. 1967. Effect of bulk density on moisture characteristics of soils. Soil Sci. 103:234–238.

26. Jackson, R. D. 1963. Porosity and salt water diffusivity relations. Soil Sci. Soc. Am. Proc. 27:123–126.

27. Klute, A. 1973. Soil water flow theory and its application in field situations. p. 9–35. In R. R. Bruce et al. (ed.) Field soil water regime. Soil Sci. Soc. Am. Spec. Pub. No. 5.

28. Laliberte, G. E., and R. H. Brooks. 1967. Hydraulic properties of disturbed soil materials affected by porosity. Soil Sci. Soc. Am. Proc. 31:451–454.

29. ————, ————, and A. T. Corey. 1968. Permeability calculated from observation data. J. Irrig. Drain Div. Proc. ASCE 94:57–71, 1968.

30. ————, A. T. Corey, and R. H. Brooks. 1966. Properties of unsaturated porous media. Hydrology Paper No. 17. Colorado State Univ.

31. Marshall, T. J. 1958. A relation between permeability and size distribution of pores. J. Soil Sci. 9:1–8.

32. Millington, R. J., and J. P. Quirk. 1961. Permeability of porous solids. Trans. Faraday Soc. 57:1200–1206.

33. Mualem, Y. 1978. Hydraulic conductivity of unsaturated porous media. Generalized macroscopic approach. Water Resour. Res. 14:325–334.

34. Philip, J. R. 1957. The physical principles of soil water movement during the irrigation cycle. Proc. 3rd Inter. Congr. Irrig. Drainage 8:125–154.

35. Reeve, M. J., P. D. Smith, and A. J. Thomasson. 1973. The effect of density on water retention properties of field soils. J. Soil Sci. 24:355–367.

36. Rose, D. A. 1963. Water movement in porous materials. Part I. Isothermal vapor transfer. Br. J. Appl. Phys. 14:256–262.

37. Shanda, A. K. 1977. Influence of soil bulk density on horizontal water infiltration. Aust. J. Soil Res. 15:83–86.

38. Tamboli, P. M., W. E. Larson, and M. Amemiya. 1964. The influence of aggregate size on soil moisture retention. Proc. Iowa Academy Sci. 7:103–108.

39. Thomas, G. W., and R. E. Phillips. 1979. Consequences of water movement in macropores. J. Environ. Qual. 8(2):149–152.
40. van Schaik, J. C., and G. E. Laliberte. 1969. Soil hydraulic properties affected by saturation technique. Can. J. Soil Sci. 49:95–102.
41. Wilkinson, G. E., and P. O. Aina. 1976. Infiltration of water into two Nigerian soils under secondary forest and subsequent arable cropping. Geoderma 15:51–59.
42. Wittmus, H. D., and A. P. Mazurak. 1958. Physical and chemical properties of soil aggregates in Brunizem soil. Soil Sci. Soc. Am. Proc. 22:1–5, 1958.
43. Yang, S. J., and E. deJong. 1971. Effect of soil water potential and bulk density on water uptake patterns and resistance to flow of water in wheat plants. Can. J. Soil Sci. 51:211–220.

Chapter 4

Tillage Effects on Soil Bulk Density and Mechanical Impedance[1]

D. K. CASSEL[2]

ABSTRACT

The degree to which various tillage operations alter soil physical properties is poorly understood and, at present, cannot be adequately predicted. In fact, few reports exist which relate short or long-term effects of tillage upon bulk density (D_b) or mechanical impedance (MI). The purposes of this review are (1) to indicate the order of magnitude of changes in D_b and MI resulting from tillage, (2) to identify major problems associated with measuring D_b and MI in field situations, and (3) to present recommendations to aid in the measurement and interpretation of D_b and MI in future tillage studies.

In field studies, D_b and MI measurements exhibit both spatial and temporal variability. The spatial variability results from vertical and lateral changes in soil properties such as texture, structure, and organic matter content and from the effects of past soil management practices. In general, each tillage operation produces non-uniform changes in soil physical properties. Hence, valid sampling to assess D_b and MI associated with specific tillage operations requires sampling as functions of depth and position (distance normal to the direction of travel of tillage tines, blades, etc.). Temporal variation in D_b and MI, if not recognized, creates problems in data interpretation.

Published D_b values from tillage studies, based upon soil core samples, range from < 1.0 to > 1.7 g/cm^3.

[1] Contribution from the Dep. of Soil Science, North Carolina State Univ. Paper No. 6684 of the Journal Series of the North Carolina Agric. Res. Service, Raleigh, NC 27650. This paper was presented at the symposium "Predicting Tillage Effects on Soil Physical Properties," December 1980 at the Am. Soc. of Agron. meeting, Detroit, MI.

[2] Professor of soil science, North Carolina State Univ., Raleigh, NC 27650.

Significant differences in D_b among tillage treatments as small as 0.07 g/cm^3 have been reported. Values of MI at in situ field capacity, as measured by the cone penetrometer and reported as the cone index (CI), range from nearly zero in a subsoil slit to values > 90 kg/cm^2 in a tillage-induced pan. Significant differences in field-measured CI as small as 1 kg/cm^2 have been reported although the pronounced variability associated with most CI measurements requires a high degree of replication.

To obtain more useful D_b and MI data in tillage studies, it is recommended that statistically sound sampling schemes for D_b and MI be developed based upon such factors as soil morphology, the geometry and physics associated with the tillage operation(s) under consideration, and temporal variation. Core sample dimensions for D_b determinations should be based upon the thickness of the soil zone to be measured and the appropriate balance between the desired level of precision and the measured D_b variability. For proper interpretation, CI data must be accompanied with the corresponding soil water content and D_b data collected at or near the point of CI measurement. Standard penetrometer designs and procedures to measure CI should be adopted by the Soil Science Society of America. When reporting D_b and CI results, the investigator must provide an adequate description of the soil(s), a description of the tillage operation(s), CI and D_b sampling details, and a summary of the statistical procedures.

INTRODUCTION

Tillage may be defined as the mechanical manipulation of soil. Some of the purposes for tillage include mixing and granulating the soil, eradicating or controlling plants, incorporating plant residues, establishing desired surface configurations, incorporating chemicals, and creating the desired degree of compactness for root growth (Gill and Vanden Berg, 1967). Many types of tillage machinery are used to accomplish these purposes. Each tillage operation, regardless of the particular tillage implement or power source, alters soil physical properties. The soil may be loosened, granulated, compacted, crushed, inverted, sheared, shattered, etc. Several of these effects may occur simultaneously. For example, moldboard plowing loosens, shears, and inverts the upper 20 cm of soil while, at the same time, the tractor wheel compacts the soil below the furrow. The tractor wheel also compacts the upper 20 cm layer ahead of the tillage tool.

One soil physical property that is nearly always altered by tillage operations is bulk density (D_b). Bulk density is easily measured and is commonly reported for both field and laboratory studies. Blake (1965) discussed several methods for measuring D_b; the most common method used by soil scientists is the core method. Typically, one or more undisturbed soil cores are collected from the desired location(s) in the field, air dried, and D_b determined. The cores are taken from either systematic or random locations on the field site. Soil core sizes range from 15 cm diam × 15 cm high (Terry et al., 1981) to rectangular cores 1 cm thick and 5 cm across (Pikul et al., 1979). More commonly, cores having diameters ranging from approximately 4 to 8 cm are collected. Additional techniques for measuring D_b include the excavation, radiation attenuation, and clod methods (Blake, 1965).

Another soil physical property modified by tillage is mechanical impedance (MI). In this paper, MI is equated to the mechanical resistance of the soil to a penetrometer. Mechanical impedance is related in a poorly understood manner to clay mineralogy and to soil physical properties such as D_b, texture, structure, water content, and percent organic matter. Tillage operations alter MI primarily by effecting changes in D_b, structure, and water content. Compared to the rather large volume of D_b data reported in the literature for tillage studies, few concise reports of MI are available.

The objectives of this review are to:

1) present published and original data to illustrate the order of magnitude of changes in D_b and MI resulting from various tillage operations,

2) identify major problems associated with the measurement of D_b and MI and the types of supporting data required for data interpretation, and

3) present recommendations to aid in the measurement and interpretation of D_b and MI data in future tillage studies.

BULK DENSITY

It is necessary to realize a priori that D_b on a particular field site varies for reasons other than the imposition of tillage operations. Lateral or horizontal variability in D_b results from changes on the landscape of such factors as soil texture, organic matter content, soil structure, and the effects of past management practices including tillage. Variation in D_b in the vertical direction is related to soil morphology with the same factors listed above being important. Furthermore, it must be recognized that soil D_b undergoes temporal variation after a tillage operation is imposed. For example, D_b of the 0 to 10-cm depth of a freshly tilled soil may increase soon thereafter due to slumping during periods of excessive wetness and to soil settling in response to desiccation and/or the kinetic energy associated with rainfall impact. As time progresses, D_b at this same depth may decrease in response to the loosening action exerted by root or animal activities.

From the outset we must acknowledge that information defining the range in D_b required for optimum plant growth is unknown for most soils. Hence, even though we may be able to measure statistically significant differences in D_b effects by tillage, the influence of this D_b change on plant growth and/or yield is not well understood. The relationship of D_b to yield has been established for relatively few soils (e.g., Flocker et al., 1960; Phillips and Kirkham, 1962; Singh et al., 1971). At less than optimum D_b, poor water relations may exist; at higher D_b, poor aeration and high MI may limit root extension.

Data Base. Bulk density of Webster silty clay loam (fine-loamy, mixed, mesic Typic Haplaquolls) was measured by Hageman and Shrader (1979) in Iowa at various depths to 90 cm following 20 years of cropping for all combinations of two crop rotations and two N fertilizer applications. Different combinations of tillage operations were imposed for the continuous corn rotation (CCCC) compared to the corn-oats [*Avena sativa* L.]-meadow-meadow rotation (COMM). Triplicate D_b

determinations at each of seven depths were measured by gamma ray transmission (Blake, 1965) in each of eight experimental plots in late summer of the year that corn was planted. No significant difference in D_b was found at the end of the 20-year period. Bulk density at the 15-cm depth for the CCCC and COMM rotations was 1.17 and 1.14 g/cm^3, respectively, and had a standard deviation of $\pm$ 0.05 g/cm^3. The investigators concluded that a large variability of D_b existed within each plot and that D_b differences, if indeed present, were masked either by original variations on the experimental site, or else that the measurement procedure used was not adequate to detect existing differences. The results of this study illustrate the importance of establishing initial or base D_b data for field plots prior to beginning a tillage study. By obtaining initial data, effects of spatial variability among plots can be minimized thus allowing more accurate interpretation of tillage-induced D_b changes for each individual plot.

Bulk density of cultivated soils following one or a series of tillage operations varies with distance normal to the direction of machinery travel. Wheel compaction associated with this traffic also varies with distances normal to the direction of travel. Moreover, additional machinery traffic on a tilled field, such as that associated with planting, spraying, etc., alters the effects of the previous tillage operation(s). The random D_b measurements of the 20-year study by Hageman and Shrader (1979) discussed above did not allow these position effects to be isolated. Position is defined as the perpendicular or normal distance from the crop row.

In order to account for positional effects of tillage, Van Diepen (1980)[3] reported D_b as a function of both depth and position. He examined the effects of four tillage treatments on D_b of a Kenansville loamy sand (loamy, siliceous, thermic Arenic Hapludults). This soil has a tillage-induced pan with an average D_b of 1.69 g/cm^3 at a depth of 30 cm. The four tillage treatments were (1) subsoiled, bedded, and planted in one operation (S-B-P); (2) chiselplowed to 30 cm and planted in one operation (CP-P); chiselplowed, disked, and planted in three separate operations (CP-D-P); and (3) conventional tillage which consisted of two diskings followed by planting (D-D-P). Soil cores of 100 cm^3 were collected at depths of 15, 30, 50, and 75 cm at four equidistant positions extending from the row to the non-trafficked inter-row.

The statistical design was a modified split-plot; it was modified because subplots (tiers) were not randomized within the whole plot. Tillage treatments were whole plots, and tiers as subplots (four depth and four position samples were taken within each tier). Hence, there were two replicates of each tillage treatment and five subplots per replicate.

Table 1 shows the analysis of variance used to test for significant differences in D_b. Although the tillage treatment effect was significant only at the 0.10 probability level, the position and depth effects, and the position $\times$ depth, treatment $\times$ position, treatment $\times$ depth, and treatment $\times$ position $\times$ depth interactions were all significant at either the 0.01 or 0.05 levels.

[3] Van Diepen, J. C. 1980. Corn response to different tillage practices in selected North Carolina soils. Unpublished M.S. Thesis. North Carolina State Univ.

Table 1. Analysis of variance of D_b of Kenansville loamy sand subjected to four tillage treatments (from Van Diepen, 1980[3]).

Source	df	MS	F Value
Replication (rep)	1	0.4023	
Tillage treatment (tmt)	3	0.1776	8.54†
Rep. × Tmt.	3	0.0208	
Tier (Rep. tmt.)	32	0.0136	
Position (pos)	3	0.0341	5.14**
Depth	3	1.3186	198.66**
Pos. × Depth	9	0.0209	3.14**
Tmt. × Pos.	9	0.0148	2.24*
Tmt. × Depth	9	0.0402	6.09**
Tmt. × Pos. × Depth	27	0.0125	1.89*
Error B	60	0.0066	
Residual	480	0.0045	
Corrected total	639	0.0142	

**,*,† Significant at the 0.01, 0.05, and 0.10 probability level, respectively.

Too often, D_b data in field studies are disregarded or wrong conclusions are drawn because the treatment main effect is not found to be significant. For most cases, tillage alters D_b only at selected depths and/or positions rather than throughout the entire profile. Hence, it follows that the main effects interactions are usually of more interest than the main effects themselves.

Figure 1A shows Van Diepen's D_b data as a function of depth in the corn row for the four tillage treatments. Significantly different values of D_b (greater than 0.07 g/cm³) as a function of position were found at both the 15 and 30-cm depths. On the other hand, no significant difference was found 32 cm from the row for either the 15 or 30-cm depth (Fig. 1B). Comparison of the data in Fig. 1A and B reveals that D_b of the S-B-P treatment increased dramatically with distance from the row and approached a value similar to those measured for the other treatments. The subsoiler was very effective in reducing D_b in the row but had little effect 32 cm from the row. Van Diepen's work shows that significant differences in D_b as small as 0.07 g/cm³ can be measured, but only if the appropriate inputs of labor, sampling scheme, and statistical analysis of the data are used.

Tiarks et al. (1974) reported data for a tillage study in which the temporal changes in D_b exceeded the differences effected by the different tillage treatments. The combined effects of tillage depth and amount of incorporated organic material upon D_b were investigated. Cattle feedlot manure at rates of 0, 90, 180, and 360 metric tons/ha/year were applied to Sharpsburg silty clay loam (fine, montmorillonitic, mesic Typic Argiudolls). Linear regression analysis of D_b measurements, taken in March 1971 (Fig. 2) showed that D_b of soil tilled to the 30-cm depth was greater than that tilled to the 10 or 20-cm depth. For the control (no manure added), the 30-cm deep tilled treatment resulted in an 0.04 g/cm³ increase in D_b. At the highest manure application rate, the maximum difference due to depth of tillage was only 0.08 g/cm³. By March 1972, D_b had decreased about 0.2 g/cm³ for all plots, including the control. This large decrease between the two dates of measurement was attributed to the effects of freezing and thawing. Comparison of data for the two dates demonstrates the importance and possible magnitude of temporal changes. For this reason, extreme caution must be exercised when analyzing D_b data collected on different dates. For the present example, the temporal change for a given treatment was approximately three times

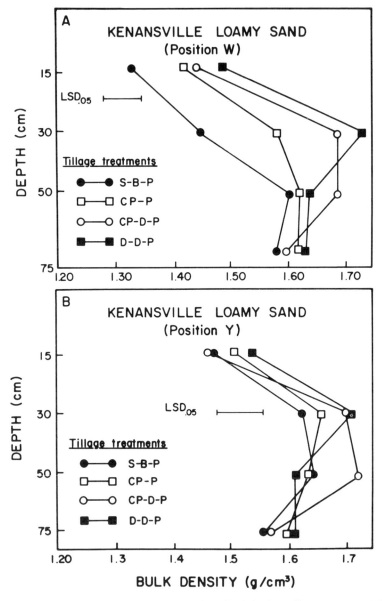

Fig. 1. Bulk density of Kenansville loamy sand vs. depth for four tillage treatments (A) in the row and (B) 32 cm from the row. Tillage treatments are defined in the text (after Van Diepen, 1980[3]).

greater than the differences among tillage treatments. In fact, it was not demonstrated that significant differences due to tillage treatments existed.

Another example of temporal changes in D_b is shown in Table 2 (D. K. Cassel, unpublished data). Tillage treatments imposed on 16 May 1976 were conventional (disked twice), broadcast chiselplowed to a depth of 27 cm at 30 cm spacings, and in-row-subsoiling to a depth of 45 cm at spacings of 95 cm plus bedding. Bulk density in the 0 to 14-cm zone for the row and nontrafficked inter-row positions increased between 16 May (planting date) and 3 June for all three tillage treatments. Each value is the mean of four 7.6 × 7.6 cm core measurements. The increase between the first two dates was attributed in this case to compaction resulting from the kinetic energy of heavy rainfall which fell on the nearly structureless soil surface. In addition to settling in response to energy from raindrop

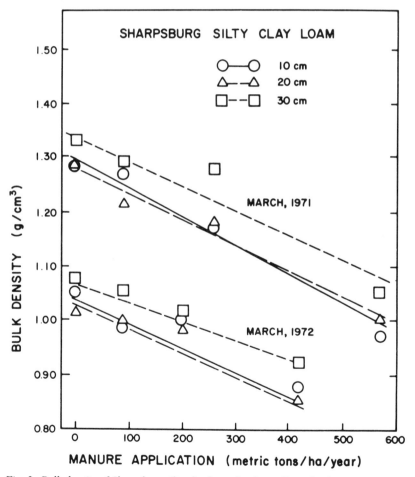

Fig. 2. Bulk density of Sharpsburg silty clay loam for three tillage depths after 1 and 2 years of manure application. Each datum is the mean of 12 measurements (after Tiarks et al., 1974).

Table 2. Bulk density of Wagram loamy sand for the 0- to 14-cm depth at two positions for three dates after imposing three tillage treatments.

	Position					
	Row			Nontrafficked inter-row		
Date	Conv.	Chisel	Subsoil	Conv.	Chisel	Subsoil
	g/cm³					
16 May	1.39±0.07	1.35±0.03	1.41±0.02	1.32±0.02	1.40±0.08	1.49±0.05
03 June	1.57±0.12	1.50±0.07	1.50±0.07	1.56±0.07	1.58±0.03	1.69±0.11
08 July	1.54±0.09	1.49±0.04	1.49±0.01	1.64±0.05	1.62±0.04	1.62±0.03
L.S.D. (0.05)	0.14	0.07	0.07	0.07	0.10	0.10

impact, some settling may also occur due to cohesive and adhesive forces during drying (Camp and Gill, 1969). No significant change occurred between 3 June and 8 July except at the nontrafficked inter-row position of the conventional treatment. The greatest measured temporal change occurred at the nontrafficked inter-row position of the conventional treatment where D_b increased from 1.32 to 1.64 g/cm³. The smallest measured temporal change over the 53-day period occurred in the row position for the subsoiled treatment and was equal to 0.08 g/cm³.

Problems Associated with D_b Measurements. Many of the problems associated with D_b sampling have been alluded to above and will only be briefly summarized. The main problem associated with D_b sampling of many past field studies is that the experimental or sampling design has been ignored. Hence, researchers have taken a few, usually too few, D_b samples from random locations in the experimental area. Usually an insufficient number of measurements were taken to perform a rigorous statistical analysis. The sampling scheme and the statistical procedures for a particular study must be developed prior to collecting samples. In developing a meaningful D_b sampling scheme, the researcher must consider the vertical and lateral patterns of effects of the tillage operation(s) so that a sufficient number of samples can be collected to isolate both depth and position effects (Van Diepen, 1980[3]). In addition, preliminary measurements must be taken to assess the degree of variability. The minimum number of samples, N, required to obtain an acceptable estimate of the mean of a population is given by

$$N = t_\alpha^2 s_e^2 / D^2 \qquad [1]$$

where t_α is the students t with (n–1) degrees of freedom (df) at the α probability level and D is the specified acceptable error. To use Eq. [1], we must first obtain an estimate of the population mean, $\bar{x}_e$, and the standard deviation, s_e^2. These values are calculated from a random sampling of D_b throughout the field. If we wish to calculate D_b to within 10% of the population mean, for example, then $D = 0.10 \, \bar{x}_e$. Examples of the above calculations are discussed by Cassel and Bauer (1975) and Terry et al. (1981).

Timeliness, i.e., the specific date(s) that D_b samples are collected is another factor that must be addressed. The time to sample depends upon the experimental objectives; it must be emphasized, however, that temporal changes occur and can affect conclusions drawn from the data. Temporal variation in D_b of freshly tilled, nontrafficked soil occurrs due to shrinking and swelling of the soil (Berndt and Coughlan, 1976); settling of soil due to cohesive and adhesive forces during drying; the action of roots, earthworms, and burrowing animals; kinetic energy associated with the impact on the soil surface of natural precipitation and irrigation water; and possibly some other factors. Special care must be taken while sampling to prevent unnecessary or accidental soil compaction.

No specific method for measuring D_b is recommended at this time. Of the methods for measuring D_b discussed by Blake (1965), the author prefers using soil cores. The nature of the study must dictate the appropriate core sample dimensions. Factors to consider in selecting core dimension are the thickness of the soil zone to be measured and the proper balance between the desired level of precision and the D_b variability known to exist in the field as determined in the preliminary sampling study.

Finally, in reporting the D_b data, it is imperative that a complete description of the soil and the tillage operation(s) be given. Details such as plowing depth, shank width and spacing, depth of disking, etc. must be reported (Rich, 1979).

MECHANICAL IMPEDANCE

Whereas D_b is the most commonly measured soil physical property in field studies, MI data are rarely reported in the literature and even fewer have been statistically evaluated. Many reasons exist for this dearth of information, some of which will be discussed in this section.

Mechanical impedance is measured with a penetrometer. Many types of penetrometers have been developed (Sanglerat, 1972) but only several types are suitable for measuring the range of MI values encountered in agricultural field soils. The cone penetrometer, which has a cone-shaped tip with a 30 to 60-degree angle, and the pocket penetrometer, which has a blunt tip, are the two types of penetrometers most often used in soils research. Details concerning their design and use are presented by Davidson (1965). During the past 10 years, soils researchers have designed and built more sophisticated penetrometers which provide a continuous record of MI vs. depth as the penetrometer is forced into the soil. Some of these penetrometers are manually operated (Carter, 1967) while others are hydraulically driven (Smith and Dumas, 1978). All MI data reported herein are reported as the cone index (CI) which is defined as the force required to push a metal cone into the soil divided by the basal area of the cone (Davidson, 1965). Cone penetrometer specifications and procedural standardization will be discussed later.

It is important that the researcher be convinced that the use of CI's is a valid method for assessing MI. Some researchers hold the viewpoint that CI measurements are of little use in field situations because field penetrometers have larger diameters than roots, they penetrate the soil faster than roots, and they are incapable of weaving in and out between soil particles like roots. Other researchers belabor the complex physics of the interaction of friction and compression forces associated with cone penetration. Still others, including the author, are cognizant of the above viewpoints but desire information concerning the hardness of soils. These researchers, although aware of interpretation problems, use the penetrometer to assess MI differences as affected by different tillage operations in a given soil.

As stated earlier, MI and hence CI are dependent upon several soil physical properties. For proper interpretation of the CI data, supporting soil physical property data must also be made available. Hence, whenever CI measurements are taken, it is also necessary to collect soil water content and bulk density data. In addition, data pertaining to soil texture and organic matter are useful, especially if CI data from different soil horizons are to be compared.

Data Base. One of the most complete sets of CI data along with supporting data of D_b and soil water content is reported by Gooderham (1976) and Gooderham and Fisher (1975). Gooderham evaluated the effect of soil water content of a silt loam soil at the time of plowing upon several soil physical properties. The soil was moldboard plowed in dry or excessively wet conditions. Tillage and cultivation of soils in excessively wet conditions is common in humid and many subhumid regions. Figure 3A shows field-measured CI's, obtained by using a 13-mm diam cone penetrometer, vs. soil depth. Each datum is the mean of 120 measurements. The mean CI was obtained by averaging 20 replicates of CI obtained on each of six sampling dates (P. T. Gooderham, personal communication, 1980). Although mean values ranged only from 10 to 18 kg/cm², significant differences in CI between the wet and dry-plowed soil were detected at the 19 and 27 cm depths.

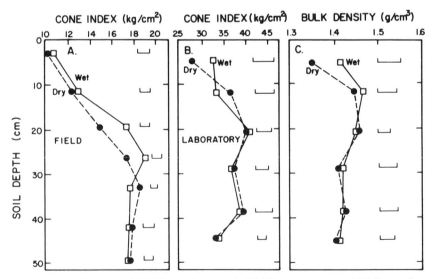

Fig. 3. (A) Field-measured CI (13-mm diam cone), (B) laboratory-measured CI (1-mm diam cone), and (C) D_b vs. soil depth for wet and dry moldboard plowed silt loam soil. The horizontal lines are L.S.D.'s for the 0.05 probability level (after Gooderham, 1976).

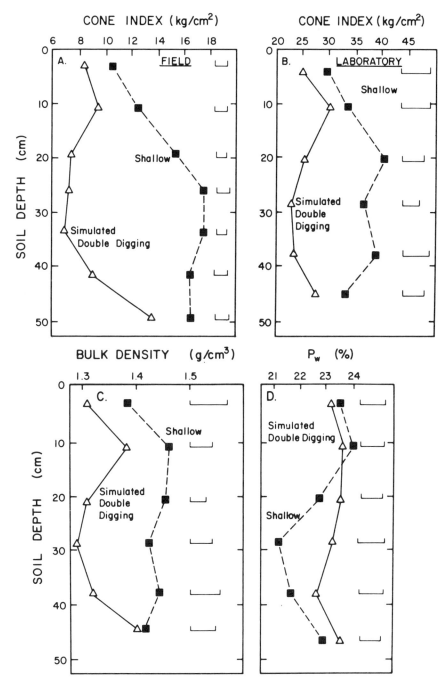

Fig. 4. (A) Field-measured CI (13-mm diam cone), (B) laboratory-measured CI (1-mm diam cone), (C) D_b, and (D) weight percent soil water vs. soil depth for simulated double digging and shallow tillage of a loam soil. Horizontal lines are the L.S.D.'s at the 0.05 probability level (after Gooderham, 1976).

One of the problems associated with the interpretation of CI measurements is illustrated by comparing Gooderham's field-measured data to laboratory-measured data on undisturbed cores of the same soil. The soil cores were collected at identical depths at the same time that the field measurements were taken. The laboratory-measured values shown in Fig. 3B are approximately two times greater than the field-measured ones. Bulk density with depth for the two treatments was not significantly different (Fig. 3C). The soil water content at corresponding depths was identical when measuring CI in the field and laboratory. The CI differences in Fig. 2A and 3B stem from differences in penetrometer geometry. A 1-mm diam penetrometer cone was used for the laboratory measurements compared to a 13-mm diam cone for the field measurements.

In addition to the above study, Gooderham (1976) also reported the results of a study to determine the effect of simulated double digging on soil physical properties. Double digging is a deep tillage operation which employs a rotovator-type operation to break up compacted soil below a moldboard plow furrow. The double digging operation was simulated in that it was performed by hand. Figure 4 shows CI, D_b, and soil water content (weight basis, P_w) for the double dug treatment (deep tilled) contrasted to the mean of four shallow-tilled treatments. None of the four shallow tilled treatments extended below the depth of mold-board plowing (30 cm). Figures 4A and B show that double digging significantly decreased CI for both field and laboratory-measured values, respectively. Cone index ranged from 7 to 17 kg/cm² for the field and from 22 to 40 kg/cm² for the laboratory-measured values. Again, the laboratory-measured CI values are approximately two times greater than the field-measured values. For this case, differences in CI stem from differences in penetrometer cone geometries, D_b (Fig. 4C), and P_w (Fig. 4D). Figure 4C shows that, in general, double digging significantly decreased D_b from values exceeding 1.4 g/cm³ to values near 1.3 g/cm³. Each D_b and P_w datum shown in Fig. 4 is the mean of 24 measurements, i.e., the mean of four measurements taken on each of six dates from the non-trafficked areas between barley [Hordeum vulgare L.) rows.

Another tillage study for which CI data is available was reported by Bishop and Grimes (1978). The 3-year study included various combinations of precision tillage (P) and conventional tillage (C) and was conducted on Wasco sandy loam (coarse-loamy, mixed, nonacid, thermic Typic Torriorthents) at Shafter, Calif. Precision tillage was defined as chiseling to a depth of 60 cm in the drill row prior to planting. Cone index was measured directly under the potato [Solanum tuberosum L.] row using a portable recording penetrometer (Carter, 1967) fitted with a 13 or 20-mm diam cone attached to a recessed 0.95 cm diam rod. Soil water content at the time of measuring CI was controlled by sprinkler irrigating all plots several days prior to measurement; hence the matric potential at the time of measurement was approximately −0.25 bar for all treatments. Figure 5 shows that CI was nearly 20 kg/cm² in the compacted plowsole zone at the 30 to 60-cm depth for the CCC plots, i.e., the control plots, which received conventional tillage for 3 consecutive years. Cone indices for the CCP and PPP treatments ranged from 1 to 6 kg/cm² in the plowsole zone, whereas they ranged from 16 to 20 kg/cm² at the corresponding depth of the PCC and CCC treatments. In addition to pronounced changes with depth, the data in Fig. 5 shows that CI underwent pronounced temporal variation as well. For this soil, the loosening of the plowsole obtained by the chiseling operation is dissipated over a 2-year period and, hence, requires chiseling at least every 2 years.

The CI data shown in Fig. 3, 4, and 5 were reported as a function of soil depth at particular positions. The CI data of Bishop and Grimes (1978) (Fig. 5) was taken in the row position whereas that of Gooderham's (1976) (Fig. 4) was taken between barley rows, i.e., the inter-row position. Depending upon the nature of the tillage operation, CI's may vary by position in a manner similar to that in which D_b varied with position as reported by Van Diepen (1980)[3]. It is again important to emphasize that each tillage operation has a specific geometry of disturbance associated with it and that CI and D_b, as well as other physical properties, are altered in both the vertical and horizontal directions.

Results of a study in which CI was measured as a function of depth and position were reported by Cassel et al. (1978). The CI's associated with three tillage operations on a Norfolk sandy loam (fine-loamy, siliceous, thermic Typic Paleudults) having a tillage-induced pan were quantified. The three tillage treatments, used in an attempt to increase the rooting depth of soybeans were: (1) moldboard plowing to 22 cm and disking

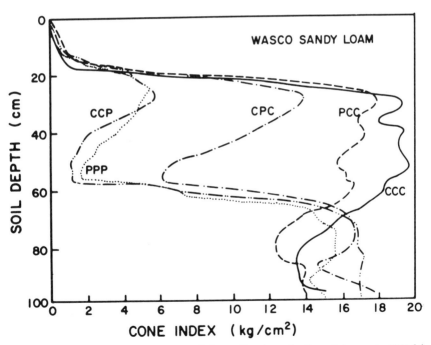

Fig. 5. Cone index profiles of Wasco sandy loam directly under the potato row in 1974 following different tillage sequences over a 3-year period. Each CI profile is the average of six replications. Matric potential was near −0.25 bar at the time of measurement; C and P refer to conventional and precision tillage, respectively. CCC signifies conventional tillage 3 years in a row (Bishop and Grimes, 1978).

twice before planting (conventional), (2) broadcast chiselplowing to 27 cm (chisel), and (3) in-row-subsoiling to 45 cm plus bedding (subsoil).

Cone index was evaluated four times during the year at seven positions and three depth increments for each of the three tillage treatments. The depths and positions for measuring CI were based upon knowledge of (1) the tillage geometry and (2) soil morphology as follows. Compaction promoted by disking occurred primarily in the upper 14 cm. Chiseling extended to a depth of 27 cm. The tillage pan occurred primarily in the 14 to 28-cm depth. Finally, it was thought that soil below 27 cm would be altered only by the subsoiling operation because only the subsoil shanks extended below this depth. In the upper 14 cm, the chiselplow and the conventional tillage operations were expected to alter the soil more or less uniformly with distance away from the row. As noted earlier, wheel traffic in the inter-row of all three treatments during planting, spraying, and harvest operations can modify the primary tillage effects. Hence, the adopted sampling design was to measure CI at three depth increments and at seven positions along a transect normal to the row. A hydraulic recording penetrometer with a 60 degree cone, an 11.1-mm diam base, and a soil penetration rate of 183 cm/min was used. The maximum CI (CI_{max}) occurring within each depth increment was analyzed using a split-split plot design with tillage as whole plot treatments arranged in four randomized blocks, positions as subplots, and depth as sub-subplots. Four sets (transects) of samples were taken per whole plot. In the analysis of variance, the variation among sets within whole plots was removed as sampling error.

Figure 6 shows CI_{max} within the 0 to 14-, 14 to 28-, and 28 to 41-cm depths vs. position for the Norfolk soil on 12 December, nearly 7 months

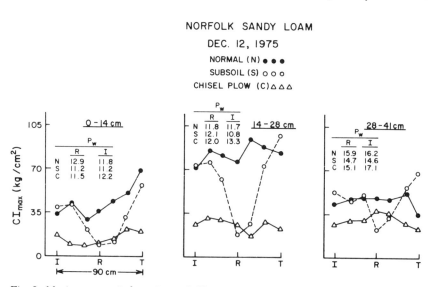

Fig. 6. Maximum cone index at in situ field capacity vs. position on 12 December for the 0 to 14, 14 to 28, and 28 to 41-cm depths. Each datum is the mean of 16 measurements. I, R, and T, refer to inter-row, row, and trafficked inter-row, respectively. The soil water content (percent by weight) at the time of CI measurement is shown for the R and I positions.

after the tillage treatments were imposed (Cassel et al., 1978). Each datum is the mean of 16 measurements. The CI_{max} values range from 15 to 90 kg/cm². The soil water content was at in situ field capacity on this date; weight percent soil water at positions R and I are also shown in Fig. 6. Maximum cone index for conventional tillage in the 0 to 14-cm depth was greatest for position T where wheel traffic occurred during planting. The greater CI_{max} values in the 14 to 28-cm depth stem from the tillage pan which has a density of 1.7 to 1.8 g/cm³. The effect of subsoiling to reduce CI_{max} was greatest at position R for all three depths. For positions I and T, CI_{max} for the subsoil treatment was nearly identical to that for conventional treatment. Maximum cone index for the chisel treatment was invariant with position but increased with depth.

Considerable variation was observed for 16 individual CI_{max} measurements used to calculate the mean values presented in Fig. 6. To gain an appreciation for this variability, selected data used to construct Fig. 6 are presented in a different manner in Fig. 7. In Fig. 7, CI (mean of 16 measurements) is shown at 5-cm intervals for the three tillage treatments

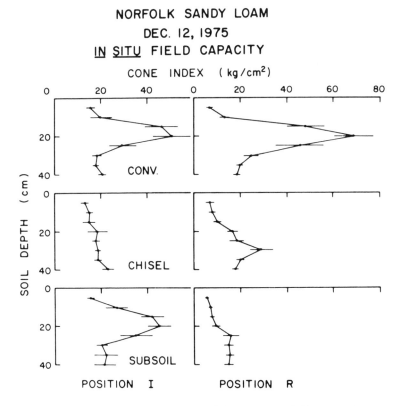

Fig. 7. Mean CI at 5-cm intervals of Norfolk sandy loam at in situ field capacity for positions I and R 7 months after imposing conventional, chisel, and subsoil tillage treatments (12 December). The horizontal line is equal to two times the standard error of the mean ± one SEM. Each datum is the mean of 16 measurements.

at positions I and R. Length of the horizontal line is equal to two times the standard error of the mean (SEM). For the conventional treatment (tillage pan present), CI exceeds 40 kg/cm² at the 15- and 20-cm depths at position I and approaches 70 kg/cm² at position R at the 20-cm depth. Chiselplowing reduced CI at both the I and R positions whereas subsoiling reduced CI only at position R. The SEM's for the CI measurements are smaller at those positions where the soil was disturbed by the tillage operations, i.e., at positions I and R for the chiselplow treatment and at position R for the subsoil treatment. Greater values of SEM are associated with the tillage pan which was not disturbed, i.e., at positions I and R for conventional tillage and at position I for the subsoil operation. Hence, fewer measurements would have been required to obtain a given level of precision for the chiselplowed treatment than for the conventional treatment.

To gain an even greater appreciation for the amount of variability associated with the field-measured CI values for this Norfolk sandy loam, Fig. 8 presents the same mean CI data for the same two positions. In this figure, the length of the horizontal line at each 5 cm depth interval is two times the standard deviation (SD). At first glance, it appears that the variation associated with the CI measurements in Fig. 8 is greater than that associated with the same data set in Fig. 7. This is not the case, of course, since the horizontal lines represent two different statistics. The consider-

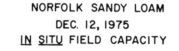

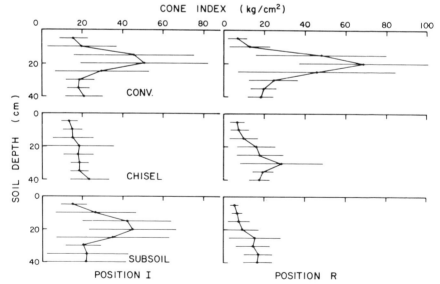

Fig. 8. Mean CI at 5-cm intervals for Norfolk sandy loam at in situ field capacity for positions I and R 7 months after imposing conventional, chisel, and subsoil tillage treatments (12 December). The horizontal line is equal to two times the standard deviation. Each datum is the mean of 16 measurements.

able variation in SD and SEM require that proper attention be given to sampling detail in order to provide sufficient replication for valid statistical analysis of the data. The CV's for the 168 treatment × depth × position combinations (three tillage treatments, eight depths, seven positions) ranged from 15 to 111% with most of the CV's falling in the 40 to 70% range.

Problems Associated with Measuring CI. Many problems exist in obtaining useful CI information in field studies. For any given soil, the procedural errors associated with making CI measurements are assumed to be systematic across tillage treatments thus allowing valid comparisons of the experimentally measured CI data. This assumption is usually justified.

One major problem alluded to earlier is that CI at any given locus in a field undergoes temporal variation. A portion of the temporal variation is directly attributable to temporal changes in D_b that were discussed previously. In addition, the soil water content undergoes temporal variation due to water additions by precipitation and irrigation and water losses by drainage, plant use, and evaporation directly from the soil. The effect of soil water suction (which is related to soil water content) and D_b on CI of Tarai loam [Mollisol] is illustrated in Fig. 9 (Singh and Ghildyal, 1977). Cone index is directly proportional to both D_b and soil water suction. At a D_b of 1.2 g/cm³, the effect of soil water suction on CI is small. However, as D_b increases, the effect of soil water suction increases dramatically. The reason for this behavior is that, as D_b increases, the soil shear strength increases because of the interlocking of soil particles. Increasing the soil water suction increases cohesion and decreases the lubricating effect of soil water. Other factors affecting CI are texture, structure, and particle surface roughness (Cruse et al., 1980).

Based upon the above discussion, it is important to select an appropriate time to measure CI. Chancellor (1976) recommends that CI measurements be taken at those times when the soil is at in situ field capacity. This soil water content is reproducible in both field soils and laboratory samples; for many soils, it is attained several days after a rainfall or irrigation which thoroughly wets soil to the desired depth of measurement. Bishop and Grimes (1978) were assured that the soil water content of Wasco sandy loam was near in situ field capacity by irrigating 2 days prior to CI measurement. For some studies, especially those without irrigation, it may be necessary to measure CI when the soil is not at in situ field capacity. This situation is likely to occur if the researcher wants to know and relate CI to particular stages of plant growth. Differences between tillage treatments may be accentuated by measuring CI when the soil is somewhat dryer than in situ field capacity. Many soils become so hard at the dryer water contents that it becomes difficult or even impossible to obtain reliable data. It is conceivable that a researcher could bias his data by indiscriminately collecting data at water contents other than in situ field capacity.

Because CI is so highly dependent upon D_b and soil water content, it is imperative that supporting D_b and water content data be collected. This supporting data, which is collected in the immediate vicinity where the CI measurement was taken, i.e., at or as near to the same depth and position as possible, allows a more comprehensive interpretation of the data than would otherwise be possible. Supporting D_b and water content information was reported for the CI data in Fig. 3 and 4 while only water content data were reported for that in Fig. 6. Failure of investigators to measure these supporting data is common because of the extensive labor requirement necessary to obtain reliable estimates of water content and particularly D_b. In addition to the supporting D_b and water content data,

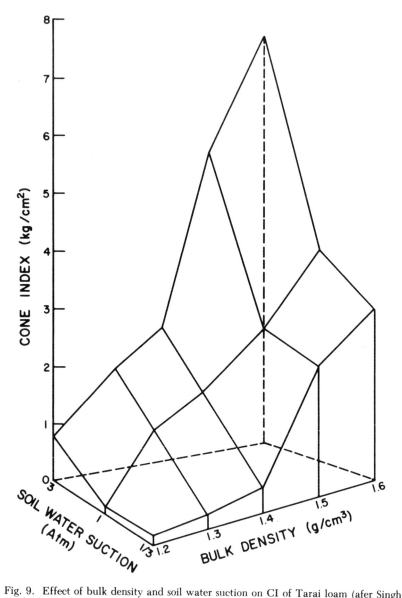

Fig. 9. Effect of bulk density and soil water suction on CI of Tarai loam (afer Singh and Ghildyal, 1977).

information concerning soil texture, percent organic matter, and in some cases, soil structure are also helpful for interpretation (Bradford, 1980).

A third problem associated with the measurement and use of CI data is the difficulty encountered in attempting to compare data obtained by one investigator for a particular soil to that obtained by a second investigator for a different soil. In fact, two investigators most likely will obtain different values for the same soil! The reason for this non-comparability of CI values is that investigators, in general, do not use identical penetrometers nor do they use identical experimental techniques. The shape of the tip (cone-shaped vs. blunt), cone angle, cone diameter, and penetration rate are factors which affect measured values (Jezequel, 1969; Gooderham, 1976; Freitag, 1967; Bowen, 1976). Freitag (1967) reported that CI for cohesive soils (clays, silts, and wet sand) is a function of penetration rate whereas that for friction soils such as dry sands is nearly independent of penetration rate. Both Freitag (1967) and Gooderham (1976) reported that smaller diameter cones give larger values compared to those obtained using larger diameter cones.

Cone penetrometers are designed so that the resistance required to push the cone into the soil flexes a proving ring which in turn causes a needle on a dial indicator to move. A calibration curve is used to convert the dial reading into CI. To minimize differences in CI measurements among investigators, the American Society of Agricultural Engineers adopted standards for penetrometer design and for the experimental techniques (ASAE, 1975). Two standard cones, each having an angle of 30 degrees, were selected. The A cone has a cross sectional area of 3.23 cm² (0.5 inches²); the B cone 1.29 cm² (0.20 inches²). The smaller cone is used in the harder soils. A penetration rate of 183 cm/min (72 inches/min) was also adopted. It is relatively easy for a researcher to adopt both the 30 degree angle and recommended cone diameters but the standard penetration rate of 183 cm/min is more difficult to obtain in field research, especially for hand-operated penetrometers. For soils having considerable CI variation with depth, it is virtually impossible for an individual to hand operate a penetrometer and sustain a constant penetration rate unless the soil is soft.

Personal experience and the work of Gifford et al. (1977) suggest that the pocket penetrometer (Soiltest, Model C1-700)[4] is useful for qualitatively detecting horizons or zones of either high or low MI in soil profiles. The instrument itself is 16 cm long, weights 230 g, and registers the force required to push the blunt, spring-loaded tip into the soil to a depth of 0.3 cm. This penetrometer has not been found too useful for quantitative studies due to difficulty sustaining a uniform penetration rate, reading the scale, and accurately determining when the tip has penetrated 0.3 cm into the soil.

Another factor that limits the usefulness of CI measurements is a poorly designed sampling scheme which will not accommodate an appropriate statistical analysis. It cannot be overemphasized that adequate

---

[4] The use of trade names in this publication does not imply endorsement by the North Carolina Agricultural Research Service of the products named nor criticism of similar ones not mentioned.

attention must be directed to the statistical design of CI sampling studies. Factors to consider in designating the sampling scheme are the time (when) to sample, the temporal and spatial variability, and the additional soil properties required to interpret the results. Because soil texture and soil structure vary spatially and because soil water content and D_b vary both spatially and temporally, one deduces a priori that CI will vary both spatially and temporally. Cone index measurements taken at random locations in a field usually fail to provide adequate information in tillage studies. As discussed earlier, the geometry of the tillage operation and soil profile properties must be analyzed prior to establishing sampling loci for a given field situation (Cassel et al., 1978). Position and depth effects are important and replicated measurements at each position × depth combination are necessary to properly assess the high degree of variability encountered in fields soils (see Fig. 6 and 7). The number of measurements required to determine CI at a predetermined acceptable error level can be computed using Eq. [1]. Data presented by Cassel and Nelson (1979) and McIntyre and Tanner (1959) suggest that penetrometer measurements may be log-normally distributed. Use of Eq. [1] is also valid for a log-normal distribution or any population distribution if N is large enough; Warrick and Nielsen (1980) have suggested that perhaps N should be as large as 20. Another statistical approach for the analysis of soil properties is the use of regionalized variables (Journel and Huijbregts, 1978).

Penetrometer use is limited to those soils devoid of gravel or rocks, the presence of which results in extremely erratic data. When a penetrometer tip encounters a small rock, the rock may be pushed aside or the direction of the penetrometer path may be deflected. If a larger rock is encountered the penetrometer shaft may be bent, especially if the penetrometer is hydraulically driven. In all cases, artificially high CI values result.

One problem rarely considered when analyzing any soil physical property data is the fact that the reference point for depth sampling changes with time as the soil surface settles (Negovelov and Unguryan, 1977). For example, the soil in the 0 to 5-cm depth soil increment immediately after disking contains fewer soil particles (and hence has a lower D_b) than contained in the 0 to 5-cm increment after settling or compaction has occurred. If there will be marked soil settling, it is reasonable to assume that structure and even soil texture at a given depth will change with time. The extent to which these changes affect CI, D_b, and other properties is unknown. A procedure may ultimately have to be developed to account for this changing reference point.

RECOMMENDATIONS

Based upon the previous discussions, the following recommendations are proposed:

BULK DENSITY

1. Develop an appropriate D_b sampling scheme. The scheme should be statistically sound and must be developed only after considering such

factors as soil morphology, the geometry and patterns of soil disturbance associated with the tillage operation(s), and expected temporal variation. If two or more tillage operations are to be compared, D_b samples must be collected at times to circumvent unwanted temporal variation.

2. Choose core sample dimensions appropriate for the thickness of the soil zone being monitored. The number of replications should be based upon the measured variability and the desired level of precision.

3. When reporting D_b data, give (a) an adequate description of the soil, (b) a description of the tillage operation(s), (c) details of the D_b sampling scheme, and (d) details concerning statistical treatment of the data.

MECHANICAL IMPEDANCE

1. Develop an appropriate sampling scheme as discussed for D_b above. The Soil Science Society of America should consider adopting several penetrometer designs and procedures for measurement of CI in both field and laboratory situations. Until this is done, it is recommended that measurements in the field be taken as often as possible using the penetrometer standards adopted by the American Society of Agricultural Engineers (ASAE, 1975).

2. Sufficiently replicated supporting D_b and water content information must be collected at or as close as possible to the loci where CI measurements are taken. In some cases, soil texture, percent organic matter, and soil structure data should also be reported.

3. Report the data using CI values. The investigator must also provide (a) an adequate soil description, (b) a description of the tillage operation(s), (c) sampling details, and (d) a summary of the statistical procedures.

SUMMARY

The degree to which various tillage operations alter soil physical properties is poorly understood and, at present, cannot be adequately predicted. In fact, few reports exist which relate short or long-term effects of tillage on bulk density (D_b) or mechanical impedance (MI). The purposes of this review were (1) to indicate the order of magnitude of changes in D_b and MI resulting from tillage, (2) to identify major problems associated with measuring D_b and MI in field situations, and (3) to present recommendations to aid in the measurement and interpretation of D_b and MI in future tillage studies.

In field studies, D_b and MI measurements exhibit both spatial and temporal variability. The spatial variability results from vertical and lateral changes in soil properties such as texture, structure, and organic matter content and from the effects of past soil management practices. In general, each tillage operation produces non-uniform changes in soil physical properties. Hence, valid sampling to assess D_b and MI associated with specific tillage operations requires sampling as functions of depth and position (distance normal to the direction of travel of tillage tines, blades, etc.). Temporal variation in D_b and MI, if not recognized, creates problems in data interpretation.

Published D_b values from tillage studies, based upon soil core samples, range from < 1.0 to > 1.7 g/cm³. Significant differences in D_b among tillage treatments as small as 0.07 g/cm³ have been reported. Values of MI at in situ field capacity as measured by the cone penetrometer and reported as the cone index (CI), range from nearly zero in a subsoil slit to values > 90 kg/cm² in a tillage-induced pan. Significant differences in field-measured CI as small as 1 kg/cm² have been reported although the pronounced variability associated with most measurements requires a high degree of replication.

To obtain more useful D_b and CI data in tillage studies, it is recommended that statistically sound sampling schemes for D_b and CI be developed based upon such factors as soil morphology, the geometry and physics associated with the tillage operation(s) under consideration, and temporal variation. Core sample dimensions for D_b determinations should be based upon the thickness of the soil zone to be measured and the appropriate balance between the desired level of precision and the measured D_b variability. For proper interpretation, CI data must be accompanied with the corresponding soil water content and D_b data collected at or near the point of CI measurement. The Soil Science Society of America should consider adoption of several penetrometer designs and procedures. When reporting D_b and CI results, the investigator must provide an adequate description of the soil(s), a description of the tillage operation(s), sampling details, and a summary of the statistical procedures.

LITERATURE CITED

1. American Society of Agricultural Engineers. 1975. Soil cone penetrometer. Recommendation R3131, Agricultural Engineers Handb. p. 368–369.

2. Berndt, R. D., and K. J. Coughlan. 1976. The nature of changes in bulk density with water content in a cracking clay. Aust. J. Soil Res. 15:27–37.

3. Bishop, J. C., and D. W. Grimes. 1978. Precision tillage effects on potato root and tuber production. Am. Potato J. 55:65–71.

4. Blake, G. R. 1965. Bulk density. In C. A. Black (ed.) Methods of soil analysis. Part 1. Agronomy 9:374–390.

5. Bowen, H. D. 1976. Correlation of penetrometer cone index with root impedance. Paper No. 76-1516. Am. Soc. Agric. Eng.

6. Bradford, J. M. 1980. The penetration resistance in a soil with well-defined structural units. Soil Sci. Soc. Am. J. 44:601–606.

7. Camp, G. R., and W. R. Gill. 1969. The effect of drying on soil strength parameters. Soil Sci. Soc. Am. Proc. 33:641–644.

8. Carter, L. M. 1967. Portable recording penetrometer measures soil strength profiles. Agric. Eng. 48:348–349.

9. Cassel, D. K., and Armand Bauer. 1975. Spatial variability in soils below depth of tillage: Bulk density and fifteen atmosphere percentage. Soil Sci. Soc. Am. Proc. 39: 247–250.

10. ————, H. D. Bowen, and L. A. Nelson. 1978. An evaluation of mechanical impedance for three tillage treatments on Norfolk sandy loam. Soil Sci. Soc. Am. J. 42: 116–120.

11. ———, and L. A. Nelson. 1979. Variability of mechanical impedance in a tilled one-hectare field of Norfolk sandy loam. Soil Sci. Soc. Am. J. 43:450–455.

12. Chancellor, W. J. 1976. Compaction of soil by agricultural equipment. Univ. Calif. Coop. Ext. Bull. 1881.

13. Cruse, R. M., D. K. Cassel, and F. G. Averette. 1980. Effect of particle surface roughness on densification of coarse textured soil. Soil Sci. Soc. Am. J. 44:692–697.

14. Davidson, D. T. 1965. Penetrometer measurements. In C. A. Black (ed.) Methods of soil analysis. Part 1. Agronomy 9:472–484.

15. Flocker, W. J., H. Timm, and J. A. Vomocil. 1960. Effect of soil compaction on tomato and potato yield. Agron. J. 52:345–348.

16. Freitag, D. R. 1967. Penetration tests for soil measurements. Paper No. 67—652. Am. Soc. Agric. Eng.

17. Gifford, G. F., R. H. Faust, and G. B. Colthorp. 1977. Measuring soil compaction on rangeland. J. Range Manage. 30:457–460.

18. Gill, W. R., and G. E. Vanden Berg. 1967. Soil dynamics in tillage and traction. U.S. Dep. Agric. Handb. no. 316. 511 p.

19. Gooderham, P. T. 1976. The effect on soil conditions of mechanized cultivation at high moisture content and of loosening by hand digging. J. Agric. Sci. 86:567–571.

20. ———, and N. M. Fisher. 1975. Experiments to determine the effect of induced soil compaction on soil physical conditions, seedling root growth and crop yield. In Soil physical condition and crop production. Ministry of Agric., Fisheries and Food Tech. Bull. 29. London.

21. Hageman, N. R., and W. D. Shrader. 1979. Effects of crop sequence and N fertilizer levels on soil bulk density. Agron. J. 71:1005–1008.

22. Jezequel, J. 1969. Les penetrometres statiques. Influence du mode d'emploi sur la resistance de pointe Bull. Liarson Lab. Routiers Ponts Chausees 36:151–160.

23. Journel, A. G., and Ch. J. Huijbregts. 1978. Mining Geostatistics. Academic Press, London.

24. McIntyre, D. S., and C. B. Tanner. 1959. Anormally distributed soil physical measurements and non-parametric statistics. Soil Sci. 88:133–137.

25. Negovelov, S. F., and V. G. Unguryan. 1977. Determination of changes in soil density after deep plowing. Pochvovedenie 1:133–138.

26. Phillips, R. E., and Don Kirkham. 1962. Soil compaction in the field and corn growth. Agron. J. 54:29–34.

27. Pikul, Jr., J. R., R. R. Allmaras, and G. E. Fischbacker. 1979. Incremental soil sampler for use in summer-fallowed soils. Soil Sci. Soc. Am. J. 43:425–427.

28. Rich, C. I. 1979. Glossary of soil science terms. Soil Sci. Soc. Am., Madison, Wis.

29. Sanglerat, G. 1972. The penetrometer and soil exploration. Elsevier Publ. Co., Amsterdam.

30. Singh, A., J. N. Singh, and S. K. Tripathi. 1971. Effect of soil compaction on the growth of soybean [Glycine max (L.) Merr.]. Indian J. Agric. Sci. 41:422–426.

31. Singh, R., and B. P. Ghildyal. 1977. Influence of soil edaphic factors and their critical limits on seedling emergence of corn (Zea mays L.). Plant and Soil 47:125–136.

32. Smith, L. A., and W. T. Dumas. 1978. A recording soil penetrometer. Trans. Am. Soc. Agric. Eng. 21:12–14, 19.

33. Terry, T. A., D. K. Cassel, and A. G. Wollum, II. 1981. Effects of soil sample size and included root and wood on bulk density in forested soils. Soil Sci. Soc. Am. J. 45:135–138.

34. Tiarks, A. E., A. P. Mazurak, and Leon Chesnin. 1974. Physical and chemical properties of soil associated with heavy applications of manure from cattle feedlots. Soil Sci. Soc. Am. Proc. 38:826–830.

35. Warrick, A. W., and D. R. Nielsen. 1980. Spatial variability of soil physical properties in the field. p. 319–344. In Daniel Hillel. Applications of soil physics. Academic Press, New York.

Chapter 5

Tillage Effects on Soil Temperature and Thermal Conductivity[1]

P. J. WIERENGA, D. R. NIELSEN, R. HORTON, AND B. KIES[2]

ABSTRACT

Soil temperature affects plant growth directly (seed germination, emergence, root growth, and nutrient uptake) and indirectly (soil water and gas flow, soil structure, and nutrient availability). Mulches, tillage, microtopography alteration, color changes, and soil water regime alteration have been used to optimize soil temperature regimes. This paper reviews some of the evidence that has been presented on qualitative and quantitative effects of minimum tillage practices on soil temperature and soil thermal properties. Further, we indicate methods that may be used to investigate the effects of tillage on soil temperature.

INTRODUCTION

One of the most important soil characteristics is temperature because of its effect on plant growth and development. Soil temperature affects seed germination, plant emergence, root growth, nutrient uptake, and plant development. Soil temperature affects plant growth indirectly

[1] Invited paper at symposium on the Effects of Tillage on Soil Physical Properties and Processes. Annual Meeting Soil Sci. Soc. of Am. Detroit, 30 Nov.–5 Dec. 1980. Journal article 860, Agric. Exp. Stn., New Mexico State University, Las Cruces, NM 88003.

[2] Professor, visiting professor, and graduate students, Dep. of Agronomy, New Mexico State Univ., Las Cruces, NM 88003.

through its effect on soil water, aeration, soil structure, nutrient availability and decomposition of plant residues. Often, soil temperature is the determining factor in plant production. Many crops cannot be grown unless the soil temperature is above a minimum level; in tropical areas, soil temperatures may be too high for growth of some crops. Furthermore, the range of optimal soil temperatures for crop production is rather narrow, compared with those of other soil physical properties.

In many areas, planting dates for crops are dependent on soil temperatures. In such areas, planting is delayed until the temperature at a given depth in the soil profile has reached a predetermined level for several consecutive days.

Many efforts have been made to change the temperature regime of soils. These include mulching, tillage, changing the color, shape and orientation of the seedbed, and changing the soil water regime. These practices were investigated and applied long before minimum tillage practices became accepted. For example, it is common knowledge that draining poorly drained soils in colder regions allows the drained soils to warm earlier in the spring, and that spraying the seedbed with asphalt can increase the soil temperature by several degrees. It should be no surprise, therefore, that minimum tillage practices, which generally result in more crop residues on the soil, also affect the temperature regime of soils. In this paper, we review some of the evidence that has been presented on the effects of minimum tillage practices on soil temperature and soil thermal properties. We also give orders of magnitude of changes in soil temperature and soil thermal properties that may be expected from minimum tillage practices. Further, we indicate methods that may be used to investigate the effects of tillage on soil temperature.

Effects of Tillage on Soil Temperature Behavior

Interpreting or evaluating soil temperature regimes is complicated by the extent and frequency of the available data. In some studies, soil temperature is measured with a simple mercury in glass thermometer and recorded once per day. In other studies, soil temperature is measured with thermocouples, often connected in parallel to give an average reading and recorded continuously with sophisticated data acquisition systems. Figure 1 shows a typical set of data that we measured with one of the latter systems. We used single thermocouples to measure temperatures several times per minute, stored the data in memory, and calculated hourly means digitally with a built-in microprocessor. Hourly means were then stored on a cassette recorder for computer analysis and plotting. The data in Fig. 1 show a rapid increase in soil temperature at the shallower depths during the morning hours. These observations imply that, if one wants to compare the effects on soil temperature of treatments such as mulching vs. no mulch, using mercury-in-glass thermometers read once per day, the readings must be taken at the same time of the day. A difference of only 1 hour in reading of the thermometers in one plot vs. another may yield significant and consistent differences in temperature not related to the treatments. However, even if all data are taken at the same time of the day, the temperature waves may be out of phase, which makes interpretation difficult. Furthermore, as a result of the strong gradient in temperature with depth in the upper soil profile, it is imperative that each temperature measurement device be located exactly at the specific depth. A deviation in placement of only 1 mm in the vertical direction in the upper 5-cm of the soil profile may have a significant effect on the final results.

Figure 2 shows the temperature variation over a 3-day period in bare soil and in a grass covered soil. There are large differences in maximum temperatures between the two locations, but smaller differences in the minimum values. Where it is impossible to measure the soil temperature continuously, one should remember that, near the soil surface, temperature taken at the time when they are near their maximum show, in general, larger differences than those taken when they are near their minimum (Chang, 1968). At lower depths, daily fluctuation in soil temperature is considerably reduced, and the time when temperatures are taken is less critical.

In row crops, the placement of the thermometer or thermocouples with respect to the plant rows is important. When the crop is only partially covering the land area, the soil between the plant rows is exposed to more direct sunlight than the soil adjacent to the plant rows. This is shown clearly in Fig. 3. Similarly, the orientation of the rows with respect to the north has important consequences on soil temperature. In the northern hemisphere, the more exposed southern side of a plant row will have a higher soil temperature. In addition, ridge-like surfaces are reported to have a higher soil temperature near the surface than level surfaces (Chirkov, 1979), due to the increase in the surface area and resultant greater adsorption of radiation. Other factors that need to be

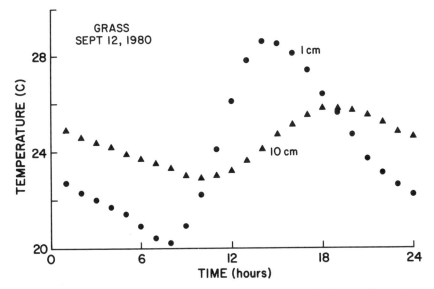

Fig. 1. Soil temperature variation at 1 and 10 cm in a grass-covered soil at Las Cruces, N.M.

taken into account in comparing soil temperature data are the size of the measurement instrument (particularly for measurement near the soil surface), the method of installation of the temperature sensors, the effects of soil disturbance during installation, conduction of heat along any lead wires, the number of measurement sites to be used, etc. Unfortunately, few studies have addressed these problems.

The effects of tillage on soil temperature have been documented in publications and some review articles (van Doren and Allmaras, 1978; Willis and Amemiya, 1973; Baeumer and Bakermans, 1973, van Duin,

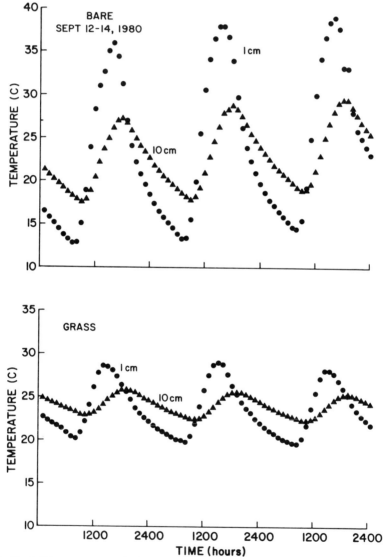

Fig. 2. Soil temperature variations at 1 and 10 cm during three clear days at Las Cruces, N.M.

1956). From these, it is clear that geographical location, type of tillage and soil cover have major effects on the response in soil temperature to tillage operations.

a. Geographical Location

Soil temperature is a function of the net amount of heat that enters or leaves the soil and of the thermal properties of the soil. The amount of heat that enters or leaves the soil surface is dependent on radiation, and the partitioning of the solar radiation in soil heat flux, convective heat and latent heat. Because solar radiation is the main energy source for warming the soil and air above it, soil temperatures vary with geographical location and generally are lower in the higher latitudes. Within a given geographical location soil temperatures also vary, because the partitioning of the energy arriving at the soil surface is dependent on the thermal properties of the soil. The color, roughness, exposure, thermal conductivity, and water content of the soil layer in contact with the air above it all have an effect on the amount of heat entering the soil. Tillage or loosening of the upper soil layer by mechanical means changes the thermal conductivity of this layer. Frequently, its color, roughness, and water content change also. Mulches, incorporated into the upper soil layer also change the thermal properties of the upper soil profile. Loosened soil or soil with mulch incorporated in the upper soil profile should warm up faster near the soil surface during periods of rising

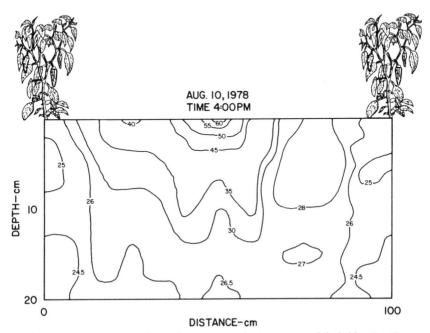

Fig. 3. Soil temperature distribution between two plant rows on a chile field in Las Cruces, N.M.

temperature, but remain cooler in the subsoil (Baeumer and Bakermans, 1973). The reverse is true during periods of falling temperature, with soils having a mulch layer in the upper profile remaining warmer in the subsoil, but cooling off at the surface.

The above-mentioned principles hold for many areas in the USA and abroad, although the magnitude of temperature changes resulting from tillage operations or mulching can vary substantially. For example, in Iowa the soil temperature at 10 cm below the soil surface covered with 3 tons per acre of grain straw mulch was reduced, on the average, by 1.0 C during May and June as compared with that of bare soil (Willis et al., 1957). Three tons per acre of grain straw depressed the mean of the daily average 10 cm soil temperature under the residue covered surface as much as 2 C compared to bare soil. A 2 C lower average soil temperature may have a significant effect on crop growth in the northern part of the USA (Allmaras et al., 1964), but may have little or no effect in areas where the soil temperature is closer to the optimum for certain crops (Unger, 1978). The magnitude of the temperature decrease resulting from tillage depends on the depth at which measurements are taken. For example, Phillips et al. (1980) reported that soil temperatures at the 2.5 cm soil depth may be decreased by as much as 6 C in the spring.

In the tropics, a decrease in soil temperature is often beneficial for crop growth. Lal (1976) reported that no tillage farming in Nigeria reduced the maximum soil temperature at 5 cm by as much as 11 C for corn, and 9 C for soybeans 2 weeks after planting.

b. Type of Tillage and Soil Cover

Different tillage systems have different effects on the soil temperature regime, because they leave different amounts of mulch on the soil surface or because they affect soil physical properties such as porosity and water content differently. Some tillage systems leave the mulch material in bands between the rows, causing a more local effect on soil temperature. Griffith et al. (1973) compared eight tillage-planting systems and found that systems that leave the most surface residue have the coolest afternoon soil temperatures. The range from lowest to highest was about 3.5 C in northern Indiana, 2.6 C in southern Indiana, and 2.0 C in eastern Indiana. Table 1 adapted from Griffith et al. (1973) shows the mean soil temperature taken in the row at 10 cm below soil surface for the 1st week after planting in 1969–1970. The data in Table 1 show the order of magnitude of changes that may be expected with the various surface treatments. Mock and Erbach (1977) found similar results.

Unger (1978) investigated the effects of various amounts of wheat straw mulch on soil temperature. He applied the wheat straw to the soil surface at rates ranging between 0 and 12 metric tons/ha, and found that increased rates decreased the average soil temperatures, maximums, minimums, and standard deviations during all seasons of fallow. The mulch effects generally were greatest in summer and spring, intermediate in fall, and lowest in winter. With 8 metric tons mulch/ha Unger (1978) measured at the 10 cm depth of a southern Great Plains soil, an average soil temperature decrease of 2.9 C in summer, 1.4 C in fall, 0.8 C in winter, and 2.3 C in the spring. As expected, maximum temperatures were affected more than mean temperatures.

The effects of plowing, rototilling, etc., on soil temperature were extensively reviewed by van Duin (1956). On the basis of theoretical calculations and field measurements he concluded that loosening the top soil reduces the heat uptake and heat loss of a soil, and causes more of the heat exchange to take place in the surface soil. As a result during periods of increasing soil temperatures, soils are warmer near the surface when tilled, and colder near the surface when left undisturbed. During periods of decreasing soil temperatures the reverse is the case. Also, the subsoil temperature in tilled soil is higher during the winter and at night, and lower during the summer and daytime than in untilled soil at identical depths. The order of magnitude of the changes, on an annual basis, is about 0.5 to 1.0 C (van Duin, 1956). Changes in soil temperature on a daily basis resulting from surface tillage operations are larger. This is shown clearly in Fig. 4 in which the amplitudes of the soil temperature at several depths below the soil surface are plotted (van Duin, 1956). The surface temperature amplitudes are considerably larger in tilled soil with a loose layer of 13 cm, than in non-tilled soil. In the subsoil the differences in amplitude between tilled and non-tilled soils are much less.

Table 1. Mean soil temperatures (C) at 1600 hours at 10 cm, in two soils planted to grain in Indiana (Griffith et al., 1973).

Tillage system	Tracy sandy loam	Runnymede loam
Conventional, spring plowed	22.4	21.7
Conventional, fall plowed	22.6	22.0
Field cultivate	22.3	21.7
Wheel track	21.6	21.7
Chisel	20.0	19.6
Till planting	21.1	20.8
Strip rotary	19.5	19.6
Coulter	18.8	18.2

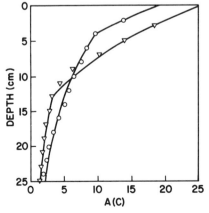

Fig. 4. Variation in amplitude of the temperature waves with depth below the soil surface for tilled (triangles) and non-tilled soil (circles) (van Duin, 1956).

The generalizations of van Duin (1956) were confirmed by Hay et al. (1978) who found that, in England, plowed soil received significantly more heat in the spring during the first 20 days after planting than direct drilled soil. Plowed soil accumulated more than twice the number of degree hours over 10 C at the 5-cm depth than did direct-drilled soil. The loosening of the surface soil layer to a depth of 2 to 4 cm decreased the mean maximum temperature of the subsoil in Northern Caucasia, USSR by 1 to 3 C, depending on the time of the year. Rolling, on the other hand, increased the mean maximum soil temperature at 3 cm by 0.6 C in April and 2.4 C in July (Chirkov, 1979).

TILLAGE EFFECTS ON SOIL THERMAL PROPERTIES

The general equation describing heat transfer in a one-dimensional isotropic medium is

$$\partial(CT)/\partial t = \partial(\lambda\, \partial T/\partial z)/\partial z \qquad [1]$$

where T is the temperature, t the time, z the depth, λ the thermal conductivity and C the volumetric heat capacity. The values of C and λ are generally not constant but vary with depth and time.

The volumetric heat capacity may be calculated with the following approximate equation (de Vries, 1963):

$$C = 0.46\, x_m + 0.60\, x_o + x_w \qquad [2]$$

where x_m, x_o, and x_w are volume fractions of soil minerals, organic matter, and water, respectively. Because x_m does not fluctuate much and x_o is usually small, it follows from Eq. [2] that the volumetric heat capacity for a given soil is nearly linearly dependent on its water content. Thus, tillage operations which affect the density and water content of a soil are also expected to affect its thermal behavior. Of course, any long term degradation of soil organic matter caused by tillage would reduce by a small extent the value of C.

The thermal conductivity may be measured (de Vries and Peck, 1958; Wierenga et al., 1969) or calculated (de Vries, 1952). The calculation procedure, first proposed by de Vries (1952), considers soil as a continuous medium of either water or air with ellipsoids of air and solids dispersed in it. He expressed the thermal conductivity of soils

$$\lambda = \sum_{i=0}^{n} k_i x_i \lambda_i / \sum_{i=0}^{n} k_i x_i \qquad [3]$$

where n is the number of different kinds of particles, x_i is the volume fraction of ith particles and λ_i is the thermal conductivity of the ith particles. The k_i values are calculated from the thermal conductivities of the ith particle and of the continuous medium, and from the shape factor of the ith particles (de Vries, 1952). The calculation procedure based on Eq. [3] was tested extensively and found to agree with measured values (de Vries,

1952, Wierenga et al., 1969; Sepaskhah and Boersma, 1979; Parikh et al., 1979). However, in some cases a correction was necessary (Skaggs and Smith, 1967; Kimball et al., 1976; Hadas, 1977). Equation [3] shows that the thermal conductivity of a soil is a function of the thermal conductivity of its constituents. Thus, practices which change the volume fractions of the constituents, (e.g., by the addition of mulch) may be expected to change the thermal conductivity of soil.

The ratio of the thermal conductivity to volumetric heat capacity is called thermal diffusivity, α. The value of α determines to some extent the rate at which a soil warms up or cools off.

It is frequently assumed that the temperature at the soil surface fluctuates sinusoidally, and that it has done so for a long time, and that the soil is semi-infinite, i.e.:

$$T(0,t) = \overline{T} + A \sin \omega t \tag{4}$$

$$\lim_{z \to \infty} T(z,t) = \overline{T} \tag{5}$$

where $\overline{T}$ is the average soil temperature, A the amplitude of the surface temperature wave, and ω the radial frequency. The solution of [1] subject to [4] and [5] is (van Wijk, 1963):

$$T(z,t) = \overline{T} + A \exp(-z\sqrt{\omega/2\alpha}) \sin (\omega t - z\sqrt{\omega/2\alpha}) \tag{6}$$

From Eq. [6] one may derive directly the amplitude equation:

$$\alpha = \omega[(z_2 - z_1)/\ln(A_1/A_2)]^2/2 \tag{7}$$

where A_1 and A_2 are the amplitudes at depths z_1 and z_2, respectively. Knowing that the maximum value of the sineterm in Eq. [6] is 1.0, one obtains the phase equation:

$$\alpha = [(z_2 - z_1)/(t_2 - t_1)]^2/2\omega \tag{8}$$

where t_1 and t_2 are the times that the maximum (or minimum) temperatures occur at depths z_1 and z_2. Equations [7] and [8] have been used successfully to obtain first order estimates of the thermal diffusivity of soils. The approach works well below 10 cm from the soil surface (Wierenga et al., 1969). Near the surface, Eq. [7] and [8] yield poorer estimates of α because the soil water content is not uniform nor is a sinusoidal surface temperature manifested (see Fig. 5). Heat transfer mechanisms other than conduction also play a role at the shallower depths (Westcot and Wierenga, 1974; Rose, 1968; Jackson, 1973).

Few investigators have determined changes in thermal properties as a result of tillage operations. Hay et al. (1978) used Eq. [7] to calculate thermal diffusivity values between 5 and 25 cm in tilled vs. non-tilled soil profiles. They calculated thermal diffusivity values for each of 25 days, and found significantly higher values in non-tilled directly drilled soil than in a plowed soil. They attributed these differences in thermal diffusivity to higher bulk densities in the direct-drilled plots, and to soil water content differences.

The lack of data on the effects of tillage on thermal properties is partially a result of incomplete measurements. Direct measurement of the thermal conductivity and volumetric heat capacity in the field is time consuming and complicated by spatial variability in soil properties, especially near the surface where the greatest interest lies. Indirect determination of the thermal diffusivity from observed temperatures fluctuations using, for example, Eqs. [7] and [8], is often impossible or difficult due to incomplete data sets. To use these equations effectively soil temperatures should be measured at least at two depths, at frequent intervals, and for several days. The observation period should include some clear days so that the assumption of a sinusoidal surface temperature wave is approximately correct. Inasmuch as the effect of tillage on soil temperature is complex and because tillage may either increase and/or decrease soil temperature depending on the depths and times of the year considered, it is obvious that soil temperatures should be measured continuously at several depths below the soil surface and for extended time periods. A comparison of temperature measurements taken at one arbitrary depth or time is of dubious value and may be misleading (Hillel, 1980). On the other hand, if measurements are made at several depths in differently treated plots, and in several replications, the task of recording and analyzing the temperature data becomes large. It is, therefore, not surprising that before the advent of magnetic tape recorders and microprocessors such studies were not frequently undertaken.

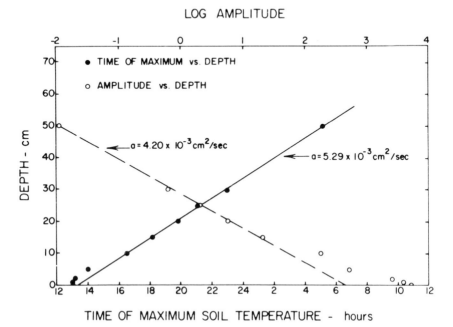

Fig. 5. Soil depth vs. logarithm of the amplitude of the soil temperature wave and versus time of maximum soil temperature (Wierenga et al., 1969).

Table 2. Values of the amplitudes and phase angles of three harmonics used to describe the soil temperature under grass and bare soil from Julian day 261–263, 1980 at Las Cruces, N.M.

Depth (cm)	Amplitudes (C)			Phase angle (rad)		
	C_1	C_2	C_3	ϕ_1	ϕ_2	ϕ_3
			Grass			
1	3.74	1.48	0.32	3.62	0.22	3.76
5	2.22	0.77	0.09	3.04	−0.64	1.86
10	1.33	0.36	0.03	2.54	−1.20	0.38
			Bare soil			
1	10.24	3.23	0.21	3.58	0.07	3.23
5	6.77	1.91	0.19	3.24	−0.46	2.05
10	4.31	1.01	0.11	2.84	−1.09	1.95

Analysis of Field Observations

For the derivation of the amplitude and phase equations (Eq. [7] and [8]) it was assumed that the temperature at the soil surface varies sinusoidal, but has a more complex form. On clear days the early morning rise is generally steeper and the afternoon decrease more gradual than a pure sine wave form. Moreover, variations in cloud cover complicate even further the temperature of the soil surface. A better description of temperature fluctuations in soil may be obtained with a Fourier expression (van Wijk, 1963; Nerpin and Chudnovskii, 1970):

$$T(t) = \overline{T} + \sum_{n=1}^{N} (A_n \cos n\omega T + B_n \sin n\omega T) \qquad [9]$$

where A_n and B_n are the Fourier coefficients, N the number of harmonics, and $\overline{T}$ the average soil temperature. Equation [9] may be written as:

$$T(t) = \overline{T} + \sum_{n=1}^{N} C_n \sin (n\omega t + \phi_n) \qquad [10]$$

where C_n is the amplitude of the nth harmonic and ϕ_n the phase angle.

Figure 1 presented soil temperature variations at 1 and 10 cm below the surface of a grass-covered field during a 3-day period with clear days. Soil temperatures were also measured at 5 cm, and in bare soil adjacent to the grass covered field. A Fourier series (Eq. [10]) with three harmonics was fitted to the observed temperature data. Values for the coefficients C_n and ϕ_n, obtained by multivariate regression, for each of the three depths are presented in Table 2. Also presented in Table 2 are the amplitudes and phase angles for three harmonics used to fit soil temperature variations observed at 1, 5, and 10 cm in the bare soil. The data in Table 2 show decreases in amplitude with depth, and also much smaller amplitudes in the grass-covered soil than in the bare soil. It is further evident that the amplitudes of the second and third harmonic are smaller than of the first harmonic. In fact, the contribution of the third harmonic to the temperature wave at 10 cm is almost negligibly small, especially for the grass-covered site. In general, the number of harmonics required to accurately describe the soil temperature over a period of time varies. Under clear skies fewer harmonics are necessary than for variable cloudiness conditions, while for greater depths also fewer harmonics are necessary because the irregularities in the temperature wave near the soil surface tend to damp out at greater depths.

Harmonic analysis has been used by some researchers to describe temperature variations in soil (van Wijk, 1963; Carson, 1963; Kalma, 1971). Carson used up to six harmonics to describe annual air and soil temperature fluctuations at Argonne, Ill., and five harmonics to describe the daily temperature wave. Results were different from 1 year to another, but from 93 to 99.8% of the total variance of the annual cycle of the soil temperature between 1 and 890 cm could be accounted for by the first harmonic. For a clear summer day the first two harmonics accounted for 99.7% of the variance. He concluded that, in general, only two harmonics are needed to describe accurately the daily cycle of soil temperatures. Kalma (1971) used two harmonics to describe the annual course of air temperature and the near-surface soil temperature. The inclusion of higher harmonics did not lead to any significant improvement in the description of the temperature waves.

Harmonic analysis may be used to describe temperature differences between tilled and non-tilled soil if sufficient observations are available. By fitting Eq. [10] to sets of temperature data one can calculate values for the coefficients and determine the effects of tillage treatments on these coefficients.

A promising approach to obtain thermal diffusivity values from the soil temperatures measured at two depths below the soil surface was developed by Horton et al. (1982). The fluctuation at the soil surface is described with a series of sine terms, e.g.:

$$T(o,t) = \overline{T} + \sum_{n=1}^{N} C_{on} \sin(n\omega t + \phi_{on}) \qquad [11]$$

where C_{on} are the amplitudes of the surface temperature waves and ϕ_{on} the phase angles of the surface temperature wave. If, further, the soil is semi-infinite and had been subjected for some time to periodic temperature changes as given in Eq. [11], a solution of [1] subject to [11] is:

$$T(z,t) = \overline{T} + \sum_{n=1}^{N} C_{on} \exp(-z\sqrt{n\omega/2\alpha}) \sin(n\omega t + \phi_{on} - z\sqrt{n\omega/2\alpha}) \qquad [12]$$

The coefficients C_{on}, ϕ_{on} and a value for $\overline{T}$ are obtained by fitting Eq. [11] to an observed data set at the upper boundary of the soil profile. By entering these coefficients in Eq. [12], the temperature at any depth and time can be calculated if α is known. If α is unknown and constant its value can be calculated by fitting Eq. [12] to a second data set measured at some depth below the first data set and by selecting values of α until the sum of squares of the differences between the measured and calculated temperatures is minimized. As an example of this procedure, Eq. [11] was fitted to soil temperatures measured at 1 cm below the surface of a grass covered plot (presented in Fig. 2) and below a bare plot. Three harmonics were used to describe the temperature at this depth. Once the parameters in Eq. [11] were determined, Eq. [12] was used to describe the temperature at a depth of 10 cm. Values for α were selected so that Eq. [12] provided the optimal fit to the soil temperature variation observed at 10 cm.

Another method of arriving at thermal diffusivity values for a particular soil is through a finite difference analysis (Wierenga et al., 1969). For homogeneous soils with constant apparent thermal diffusivity, Eq. [1] can be approximated by the following explicit finite difference equation:

$$(T_j^{n+1} - T_j^n)/\Delta t = \alpha \, (T_{j+1}^n - 2T_j^n + T_{j-1}^n)/(\Delta z)^2 \qquad [13]$$

where j is the depth interval, n the time interval, and the other symbols as previously defined. Now assume that the soil temperature was measured at frequent time intervals at two depths below the soil surface, and assume further that an initial temperature distribution with depth was known (precise knowledge of the initial temperature distribution is not critical; a good guess, or a linear interpolation between the two given temperatures will suffice). Equation [13] is used directly to evaluate the soil-temperature fluctuation at any depth between the two measured temperature fluctuations for a given value of the thermal diffusivity, because the only unknown in Eq. [13] is T_j^{n+1}. By comparing the predicted temperature fluctuation at this intermediate depth with a measured temperature variation at this depth, the correct thermal diffusivity is then obtained by minimizing the error between estimated and calculated temperature variations at the intermediate depth. This procedure can of course be used to compare observed and predicted values at several intermediate depths. No assumptions are made concerning the shape of the surface temperature wave or concerning periodicity and therefore this procedure has wide applicability. However, it is assumed that the soil is uniform with time and depth. Furthermore Eq. [13] is numerically stable only if:

$$2\alpha\Delta t/(\Delta z)^2 < 1 \qquad [14]$$

This condition can easily be met by taking relatively small time steps in the calculations (e.g., on the order of 30 min for $\Delta z = 5$ cm, 5 min for $\Delta z = 5$ cm, 4 min for $\Delta z = 2$, and 1 min for $\Delta z = 1$ cm, assuming $\alpha = 0.005$ cm^2/sec). Equation [13] was preferred above other numerical schemes which are unconditionally stable (Hanks et al., 1971), because it is easy to program and can readily be solved on small desk computers. Where soil temperatures are not recorded at short time intervals, temperatures can be generated at intermediate times by harmonic analysis as shown before, by cubic spline methods (Erh, 1972; Kimball, 1974) or by non-linear function generators (Wierenga and deWit, 1970).

Table 3 presents thermal diffusivity values for the grass-covered and bare soil for which the temperature fluctuations were shown in Fig. 1 and

Table 3. Thermal diffusivity (X 10^{-3} cm^2/sec), Days 261–263, 1980.

Depth (cm)	Grass		Bare soil	
	1–10	10–50	1–10	10–35
Amplitude equation: Eq. [7]:	2.6	6.0	4.4	5.4
Phase equation: Eq. [8]:	2.2	5.4	8.9	5.7
Fourier analysis: Eq. [12]:	2.7	5.6	5.1	6.1
Numerical analysis: Eq. [13]:	1.8	6.5	5.8	5.8
Average:	2.3	5.9	6.0	5.8

2. Diffusivity values were computed with the amplitude and phase equations (Eq. [7] and [8]), by Fourier analysis (Eq. [12]) and with the numerical procedure (Eq. [13]). Computations were made for the surface soil (1 to 10 cm), and for the subsoil (40 to 50 cm for grass-covered soil, and 10 to 35 cm for the bare soil). For all methods the soil is assumed to be uniform over the depth considered and only heat transfer by conduction is considered. The data show considerable differences in thermal diffusivity values obtained with the four procedures, even though the analysis was applied to data from three clear and sunny days with smooth temperature fluctuations. For example, the thermal diffusivity computed for the 1 to 10 cm layer in the bare soil from the phase equation was double its value computed from the amplitude equation. Because no assumptions were made concerning the shape of the surface temperature wave for the Fourier analysis and the numerical method, these methods should be more reliable.

The differences in the values obtained with the four methods demonstrate the difficulties that are encountered when analyzing soil temperature data. On the other hand, with either the Fourier analysis or the numerical approach it should be possible to determine changes in soil thermal diffusivities as a result of tillage practices. For example, the data in Table 3 clearly show a much lower thermal diffusivity value for the 1 to 10 cm layer of the grass-covered soil, possibly as a result of drier surface soil conditions or organic matter accumulations. More detailed analyses of soil temperature data will have to be made to fully understand the effects of tillage on soil thermal properties, and to be able to predict the effects of tillage on soil temperature behavior.

Spatial Variations

Although most investigators are aware of the variability in space of the temperature of a soil or its variability in thermal properties, few efforts have been made to define this variability. One notable exception is the study by Scharringa (1976). This investigator measured soil temperatures at 25 points in a 4 × r m grid system. At each of the grid points he measured soil temperatures at the 5, 10, 20, 50, and 100 cm depths, three times a day, e.g., 0800, 1300, and 1700 hours CET. Measurements were made with thermocouples in a carefully leveled and prepared field plot at De Bilt, The Netherlands. Scharringa found that the standard deviation of the mean varied with time of the year, time of the day, and depth below the soil surface. During the summer months, the standard deviation of the mean temperature at 5 cm was as high as 1.35 C. Standard deviations for the 5 cm depth were generally lower at 0800 and 1700 hours than at 1300 hours CET. At a depth of 100 cm standard deviations ranged from 0.23 to 0.68 C, and did not vary much with time of the day, but appeared somewhat higher in the summer months than in the winter months. Under cropped conditions, the variance in soil temperature at 5 cm was, as expected, strongly dependent upon degree of crop cover, and the condition of the crop (wilted vs. non-wilted).

Aguirre Luna (1979)[3] measured the soil temperature at 5 depths and 5 locations in a plane perpendicular to the plant rows. The temperatures were measured with single thermocouples. An example of the soil temperatures observed by Aguirre Luna at 1600 hours on 10 Aug. 1978, under furrows planted to chile peppers is given in Fig. 6. It is readily apparent that there is considerable variability in the measured values and that the soil temperatures along any horizontal plane within the soil profile are not the same. At the soil surface between the plant rows, the temperature is more than 30 C higher than values measured under the rows near the soil surface. Hence, at this level of observation, we observe differences in temperatures both horizontally and vertically that manifest heat conduction in several directions.

---

[3] Aguirre, Luna O. 1979. Observed and simulated 2-D soil temperature distributions under trickle-irrigated chile. Unpubl. Ph.D. Thesis. New Mexico State Univ., Las Cruces.

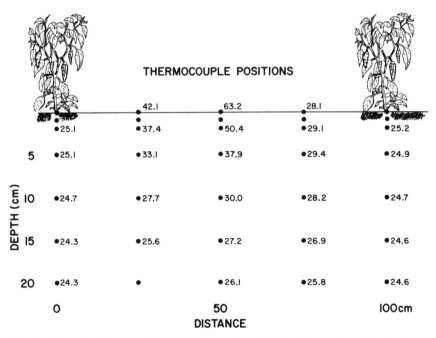

Fig. 6. Schematic diagram of thermocouple positions (dots) within a soil profile relative to crop row spacing. The plant height on 10 Aug. 1978 was about 60 cm.

Contour lines of equal temperature can be drawn through the data points using a kriging process that is based upon the spatial variance structure of the measurements (Warrick and Nielsen, 1980). The variance structure is examined with a variogram $\gamma(h)$ calculated from

$$\gamma(h) = \{ \sum_{i=1}^{N(h)} [T(X_i + h) - T(X_i)]^2 \}/2N(h) \qquad [14]$$

where X_i is the ith location of a temperature measurement, and h is the distance between any two pairs of measurement and N is the number of pairs of measurement. For the data of Fig. 6, the variogram given in Fig. 7 shows that the difference in temperature for any pair of measurements linearly depends upon the separation distance for values $0 \leq h \leq 18$ cm. For distances greater than 18 cm, the measurements are spatially independent. Values of soil temperature T_0 at locations intermediate to those measured can be estimated by T_0^*

$$T_0^* = \beta_1 T(X_1) + \beta_2 T(X_2) \cdots + \beta_n T(X_n) \qquad [15]$$

where values of the weight funtions β_i for the n locations X_i are calculated from $\gamma(h)$ subject to:

$$E(T_0^* - T_0) = 0 \qquad [16]$$

and

$$\mathrm{var}\,(T_0^* - T_0) = \min \qquad [17]$$

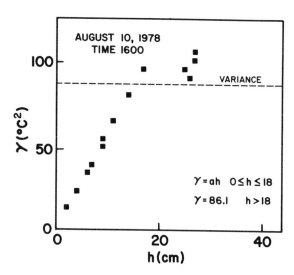

Fig. 7. Variogram of soil temperatures recorded at thermocouple positions shown in Fig. 6.

The contours of equal temperature given in Fig. 3 were drawn as a result of kriging 2086 additional values from the original 35 shown in Fig. 6. Owing to the fact that the chile pepper rows were oriented approximately in the east-west direction, the higher soil surface temperatures at the left, and the lower values at the right are a result of direct sun exposure and the shading by the southern crop row, respectively. At 1600 hours, the contours indicate that heat was moving laterally from the center to underneath the plants as well as generally downward. Obviously further development of the canopy would continue to modify the heat transfer within the profile. Any tillage operation that would alter the geometric configuration or direction of the crop row, or would give rise to an uneven thickness of cultivated topsoil would also generate temperature regimes having different levels of horizontal heat transfer.

Earlier in the morning (0600 hours) on the same day the soil temperature distribution within the profile was much different from that shown in Fig. 3. Although there is still some horizontal transfer (to the center near the soil surface, and away from the center at depths greater than 5 cm), the bulk of the heat transfer is occurring vertically upwards in the morning (Fig. 8). Thermal gradients calculated from the contours given in Fig. 3 or 8 coupled with estimates of the thermal conductivity can be used to ascertain quantitative measures of the heat flux. From each of the figures it is readily apparent that heat flux plates or other sensors positioned horizontally will not necessarily provide accurate estimates of the heat balance.

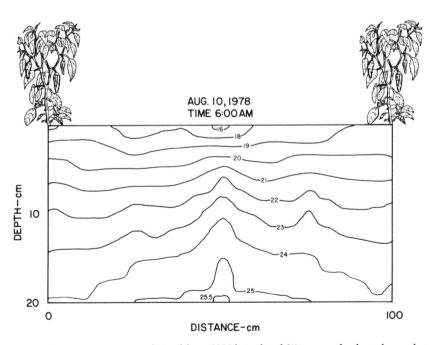

Fig. 8. Temperature contours derived from 2086 kriged and 35 measured values observed at 0600 hours 10 Aug. 1978.

To accurately assess the heat flux into or out of soil with a row crop would require a large number of heat flux plates, particularly during the early growth stages when the crop is not fully covering the soil. An alternate method would be to take soil temperature measurements (as in Fig. 6), krige soil temperatures in a detailed grid in a plane perpendicular to the crop rows and from knowledge of the heat capacity at each grid point calculate the heat flux with the standard integral method. The calculations result in vertical and horizontal soil heat flux densities in the plane perpendicular to the rows.

The depth and horizontal spacings at which temperatures were measured (Fig. 6) were somewhat arbitrary. Whether or not a greater or lesser number of observations should be taken, and at what locations they should be made can be answered by examining the uncertainty of the kriged values in terms of the kriging variance σ^2

$$\sigma^2 = \sum_{i=1}^{n} \beta_i \gamma(X_i - X_o) + \mu \qquad [18]$$

where X_o is the location of the kriged value and μ is the Lagrange multiplier. Contours of equal kriging variances applicable for Fig. 3 and 8 are given in Fig. 9 and 10, respectively. For the afternoon temperatures, the variances midway between the columns of sensors (at distances of 12.5, 37.5 cm, etc.) are as large as 100 C^2 while those in the early morning are only about 16 C^2. The upper limit of the variances would have to be chosen relative to the objective of a particular set of observations.

The above kriging analysis could also be done for data collected at the same position as a function of time instead of for observations at the same time as a function of location. Moreover, the sampling locations could be taken across an entire field as well as within the soil profile to

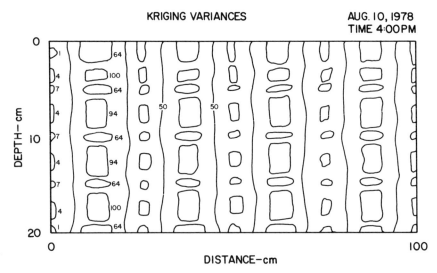

Fig. 9. Contours of kriging variances associated with the temperature contours in Fig. 3.

whatever depth is required. Inasmuch as soil temperatures together with thermal conductivity and diffusivity values are related to other soil properties affected by tillage such as bulk density, penetrability, etc., it is also possible to use co-kriging techniques (Journel and Huijbregts, 1978; Davis, 1973) to better ascertain the impact of different tillage methods. Co-kriging involves the construction of variograms for two sets of observations (e.g., soil temperature and soil penetrometer readings). Over-flight observations of soil temperature being made more frequently, at smaller distance intervals, and at less cost could be used to determine the distribution of more expensive penetrometer readings across a field through co-kriging.

The spatially repetitive nature of tillage operations may potentially give spatially repetitive values of soil physical properties including temperature regimes. A few random observations would perhaps not indicate such cyclic nature associated with tillage. Using a larger number of observations that would transcend furrows, borders, etc., as well as pedalogic soil mapping units, it is possible to identify such features using spatial cross-correlation techniques. For equally spaced observations of T and soil bulk density D along a transect across a field the cross-correlation coefficient r is calculated from

$$r(m) = cov(T,D)/\partial_t \partial_d \qquad [19]$$

where $cov(T,D)$ is the covariance of the overlapped portions of the two sequences, and s_t and s_d are the standard deviations of the overlapped portions of the two sequences. The value of r plotted against the value of m (the index of the extent and position of the overlap) is a correlogram. An examination of the correlogram identifies similarities between the spatial

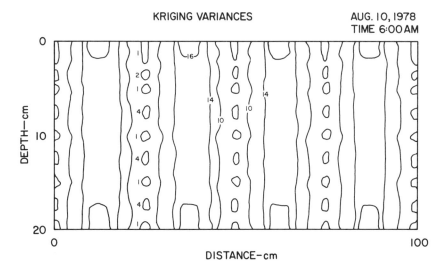

Fig. 10. Contours of kriging variances associated with the temperature contours in Fig. 8.

variations of the two parameters T and D, and hence can be used to understand and explain spatial patterns of soil physical properties associated with tillage operations.

SUMMARY AND FUTURE OUTLOOK

Observations of soil temperature provide an excellent means for ascertaining the effects of tillage on the physical condition of the soil profile as regards crop production. Such observations must nonetheless be taken at time and space intervals appropriately selected to provide meaningful interpretation. Owing to the extremely dynamic nature of the heat transfer process, interpreting soil temperature data requires a thorough understanding of the process coupled with an appreciation of the spatial variability of the soil and its crop canopy. Should many observations at one soil depth or half that many at two soil depths be taken? The appropriate answer must be developed for the particular objective and conditions associated with monitoring soil temperature.

Different tillage operations and sequences are known to create a myriad of soil conditions even on the same field that are difficult to describe quantitatively or to assess from one field to another. A quantitative assessment of the impact of tillage on the thermal properties of soil is both feasible and practical with properly executed soil temperature monitoring. Even small, yet significant changes in α can be ascertained easily with the methods presented. Moreover, a knowledge of their potential changes could indeed be one additional criterion for the selection of a particular tillage operation.

A change in the thermal properties of soil, particularly the topsoil, modifies the magnitude of the amplitude of the daily temperature wave. It should be noted that the amplitude can be altered, while the mean temperature throughout the profile may remain essentially invariant as regards the response of a crop. On the other hand, a brief increase or decrease in the amplitude may cause a significant physiological change in the crop as well as an eventual change in its yield. This is particularly true for young seedlings having roots that grow progressively deeper into the profile. Under those conditions one cannot expect a simple relationship between root development and soil temperatures. Besides the impact of temperature changes on crops created by tillage, it is important to note that chemical, physical, and microbiological soil processes are also altered.

In the future we expect greater emphasis on soil temperature and particularly that near the soil surface. In the above considerations, it was tacitly assumed that the soil surface temperature was known. Rather than use thermocouples or other sensors embedded in the soil surface, aboveground radiation observations will be more prevalent with analyses of spatial and temporal variations in the values a necessity. Variations in color, relief, albedo, water content, etc., created in part by tillage will be monitored by overhead observations. And, with the thermal properties of the topsoil dominating the temperatures of the entire profile, we anticipate an increasing effort to quantify it on a more frequent basis than in the past. In some cases, daily or weekly assessments may prove fruitful when the topsoil condition transforms rather rapidly following a particular tillage operation. We also believe that geostatistical analyses will become commonplace, with methods similar to those introduced here being used to ascertain other important soil and crop parameters related to soil temperature but too expensive to monitor over large land areas.

LITERATURE CITED

1. Allmaras, R. R., W. C. Burrows, and W. E. Larson. 1964. Early growth of corn as affected by soil temperature. Soil Sci. Soc. Am. Proc. 28:271–275.

2. Baeumer, K., and W. A. P. Bakermans. 1973. Zero tillage. Adv. Agron. 77–120.

3. Carson, J. E. 1963. Analysis of soil and air temperature by Fourier techniques. J. Geophys. Res. 68:2217–2232.

4. Chang, Jen-Hu. 1968. Climate and agriculture. Aldine Publ. Co., Chicago. p. 304.

5. Chirkov, Y. I. 1979. Soil climate. p. 325. In J. Seemann, Y. I. Chirhov, J. Lomas, and B. Primault (ed.) Agrometeorology. Springer-Verlag, Berlin, N.Y.

6. Davis, J. C. 1973. Statistics and data analysis in geology. John Wiley & Sons, Inc., New York. 550 p.

7. de Vries, D. A. 1952. The thermal conductivity of soil. Meded. Landbouwhogesch., Wageningen. 72 p.

8. ————. 1963. Thermal properties of soils. p. 210–235. In W. R. van Wijk (ed.) Physics of plant environment. North-Holland Publishing Co., Amsterdam.

9. ————, and A. J. Peck. 1958. On the cylindrical probe method of measuring thermal conductivity with special reference to soils. I. Extension of theory and discussion of probe characteristics. Aust. J. Phys. 11:255–271.

10. Erh, K. T. 1972. Application of the spline function to soil science. Soil Sci. 114:333–338.

11. Griffith, D. R., J. V. Mannering, H. M. Galloway, S. D. Parsons, and C. B. Richey. 1973. Effect of eight tillage-planting systems on soil temperature, percent sand, plant growth, and yield of corn on five Indiana soils. Agron. J. 15:321–326.

12. Hay, R. K. M., J. C. Holmes, and E. A. Hunter. 1978. The effects of tillage, direct drilling and nitrogen fertilizer on soil temperature under a barley crop. J. Soil Sci. 29: 174–183.

12. Hadas, A. 1977. Evaluation of theoretically predicted thermal conductivities of soils under field and laboratory conditions. Soil Sci. Soc. Am. J. 41:460–466.

13. Hanks, R. J., D. D. Austin, and W. T. Ondrechen. 1971. Soil temperature estimation by a numerical method. Soil Sci. Soc. Am. Proc. 35:655–667.

14. Hay, R. K. M., J. C. Holmes, and E. A. Hunter. 1978. The effects of tillage, direct drilling and nitrogen fertilizer on soil temperature under a barley crop. J. Soil Sci. 29:174–183.

15. Hillel, D. 1980. Fundamentals of soil physics. Academic Press, N.Y. p. 413.

16. Horton, R., P. J. Wierenga, and D. R. Nielsen. 1981. Determining the apparent thermal diffusivity of soil near its surface (submitted for publication). SSSAJ.

17. Jackson, R. D. 1973. Diurnal changes in soil-water content during drying. p. 37–56. In R. R. Bruce K. W. Flach and H. M. Taylor (ed.) Field soil water regime. Soil Sci. Soc. Am. Spec. Publ. no. 5.

18. Journel, A. G., and Ch. J. Huijbregts. 1978. Mining geostatistics. Academic Press, New York. 600 p.

19. Kalma, J. D. 1971. The annual course of air temperature and near surface soil temperature in a tropical savannah environment. Agric. Meteorol. 8:293–303.

20. Kimball, B. A. 1974. Smoothing data with Cubic Splines. Agron. J. 68:126–129.

21. ————, R. D. Jackson, R. J. Reginato, F. S. Nakayama, and S. B. Idso. 1976. Comparison of field-measured and calculated soil heat fluxes. Soil Sci. Soc. Am. J. 40:18–25.

22. Lal, R. 1976. No-tillage effects on soil properties under different crops in western Nigeria. Soil Sci. Soc. Am. J. 40:762–768.

23. Mock, J. J., and D. C. Erbach. 1977. Influence of conservation-tillage environments on growth and productivity of corn. Agron. J. 69:337–340.

24. Nerpin, S. V., and A. F. Chudnovskii. 1970. Physics of the soil. Israel Program for Scientific Translations. Keter Press, Jerusalem. p. 466.

25. Parikh, R. J., J. A. Havens, and H. D. Scott. 1979. Thermal diffusivity and conductivity of moist porous media. Soil Sci. Soc. Am. J. 43:1050–1052.

26. Phillips, R. E., R. L. Blevins, G. W. Thomas, W. W. Frye, and S. H. Phillips. 1980. No-tillage agriculture. Science 208:1108–1113.

27. Rose, C. W. 1968. Water transport in soil with a daily temperature wave. I. Theory and experiment. Aust. J. Soil Res. 6:31–44.

28. Scharringa, M. 1976. On the representativeness of soil temperature measurements. Agric. Meteorol. 16:263–276.

29. Sepaskhah, A. R., and L. Boersma. 1979. Thermal conductivity of soils as a function of temperature and water content. Soil Sci. Soc. Am. J. 43:439–444.

30. Skaggs, R. W., and E. M. Smith. 1967. Apparent thermal conductivity of soil as related to soil porosity. Paper No. 67-114 presented at the Annual Meeting of the ASAE at Saskatoon, Saskatchewan, 27–30 June.

31. Unger, P. W. 1978. Straw mulch effects on soil temperatures and sorghum germination and growth. Agron. J. 70:858–864.

32. van Doren, D. M., and R. R. Allmaras. 1978. Effect of residue management practices on the soil physical environment, microclimate, and plant growth. p. 49–84. In Am. Soc. Agron. Spec. Pub. no. 31. Crop residue management systems. ASA, CSSA, SSSA, Madison, Wis.

33. van Duin, R. H. A. 1956. On the influence of tillage on conduction of heat, diffusion of air and infiltration of water in soil. Versl. Landbouwkd. Onderz. no. 62. 7:82.

34. van Wijk, W. R. 1963. Physics of plant environment. North-Holland Publishing Co., Amsterdam.

35. Warrick, A. W., and D. R. Nielsen. 1980. Spatial variability of soil physical properties in the field. p. 386. In D. Hillel. Applications of Soil Physics. Academic Press, N.Y.

36. Westcot, D. W., and P. J. Wierenga. 1974. Transfer of heat by conduction and vapor movement in a closed soil system. Soil Sci. Soc. Am. Proc. 38:9–14.

37. Wierenga, P. J., and C. T. deWit. 1970. Simulation of heat transfer in soils. Soil Sci. Soc. Am. Proc. 34:845–848.

38. ————, D. R. Nielsen, and R. M. Hagen. 1969. Thermal properties of a soil based upon field and laboratory measurements. Soil Sci. Soc. Am. Proc. 33:354–360.

39. Willis, W. O., and M. Amemiya. 1973. Tillage management principles: soil temperature effects. p. 22–42. In Conservation tillage. Proc. of a Natl. Conf. Soil Conserv. Soc. Am. Iowa. p. 22–42.

40. ————, W. E. Larson, and D. Kirkham. 1957. Corn growth as affected by soil temperature and mulch. Agron. J. 49:323–328.

Chapter 6

Tillage Effects on Soil Aeration[1]

A. E. ERICKSON[2]

ABSTRACT

Tillage can have a profound effect on the aeration conditions of soil depending on the initial condition of the soil and the subsequent rainfall. Of the aeration measurements available, oxygen diffusion rates best describe the soil-plant interactions. Because the aeration effects are transient, the measurements must be performed punctually during the period of aeration stress. Data from aeration studies can be very useful in describing the beneficial effects of tillage on plant growth and yield.

INTRODUCTION

Tillage can have a profound effect on the aeration of the soil with the magnitude of the change depending on the initial soil properties. Tillage of a dense soil with poor aeration characteristics with the proper tool at the right moisture condition can correct temporarily the aeration problem. A coarse textured or well aggregated soil without an aeration problem cannot be improved by tillage and is a prime candidate for no-till agriculture. What Burnett and Hauser wrote in 1968 about deep tillage is also true of all tillage. "Deep tillage is not a panacea for all crop production problems associated with physical or chemical soil properties. Where

[1] Contribution from the Michigan Agric. Exp. Stn., East Lansing, MI 48824. AES J. Ser. No. 10122.

[2] Professor of soil science, Crop and Soil Sciences Dep., Michigan State University.

Table 1. Air capacity* of Brookston clay as affected by tillage.

Plot treatment	Air capacity, 0 to 7.5 cm deep percent by volume	
	11 June 1940	25 Sept. 1940
Fall-plowed 1939, plot left flat	10.3	3.5
Fall-plowed 1939, plot ridged	15.3	11.8
Spring-plowed 1940, plot left flat	15.9	3.9

* Noncapillary porosity was originally used for air capacity. From Baver and Farnsworth, 1940.

certain soil factors that limit plant growth can be identified deep tillage may be beneficial."

Whether plants respond to the changes brought about by tillage depends on the subsequent water regime. During wet years, or after intense rains or irrigation the oxygen supply in the soil is inadequate for the plant roots and beneficial effects of tillage may be realized. Conversely, in dry years there may be no benefits accruing to tillage. At the Ferden Experiment Farm in Michigan beneficial effects of proper soil management occurred only about 1 year in 4. On soils with more acute physical problems, the frequency is higher.

The dynamics of aeration complicates the problem because in most situations aeration is a problem only during wet periods.

The proper tillage manipulation is important in correcting the problem. For example, Kalamazoo sandy loam (a fine-loamy, mixed, mesic Typic Hapludalfs) has a very dense subsurface horizon at 20 to 35 cm deep, which when wet is a barrier to root development due to inadequate aeration but physically impedes roots when dry (Piper[3]). Deep chiseling of this soil, when it is dry, shatters the dense horizon and alleviates the condition. When the soil is rewetted, which occurs each winter, it returns to its original condition. Deep mixing with a giant disc plow distributed chunks of the dense horizon into the sandy loam surface soil and corrected the problem.

If tillage is used to correct aeration problems, soil aeration measurements could assist in the development of the soil tillage practices (Cannell, 1977). Grable (1968) has a review of the subject and Cannell (1977) also addresses the subject. There are, however, relatively few tillage experiments which have been critically evaluated with aeration measurements.

There is another growing body of literature which considers soil aeration and its relation to plants, Russell (1952), Vilain (1963), Pearson (1965), Grable (1966), Stolzy (1974), Cannell (1977), Smith (1977), and Armstrong (1979) have all reviewed this literature. Much of this research has been done in the laboratory or greenhouse, but has application to field conditions where tillage is performed. There are new methods and instruments to measure aeration parameters in the field, Raney (1949), Lemon and Erickson (1952), van Bavel (1954), Willey and Tanner (1963), Letey and Stolzy (1964), Tacket (1968), Phene et al. (1976), and Patrick (1977).

---

[3] Piper, C. D. The influence of deep mixing a Kalamazoo sandy loam on several physical factors and corn root development and yield. Doctor's Dissertation. Michigan State University (1967).

The critical limits of these parameters in relation to plants are becoming better defined. Theoretical studies have pointed to the crucial parameters. Tillage experiments can now be monitored fairly well to determine the adequacy of soil aeration in terms important to the production of crop plants.

Aeration measurements can be divided into the kind of parameter measured; capacity, intensity, or rate. Capacity has been called air porosity or non-cappilary porosity and is the volume fraction occupied by gas. Intensity is the oxygen concentration in the gas-filled pores or oxygen dissolved in the soil solution. The rate factor is the diffusion rate either in the gas-filled pores or through the liquid films.

Air Capacity

Air capacity, or the soil volume fraction occupied by gas, is the parameter that has been most used to describe aeration. Often it is designated as the air volume associated with water at some water potential that seems reasonable or at the formerly respectable field capacity. One reason for the popularity of air capacity is that it can be calculated from the soil moisture characteristic.

The data of Baver and Farnsworth (1940) illustrate the use of air capacity or non-capillary pore space in tillage research. Table 1 shows the differences produced in the soils due to tillage treatments. Figure 1 shows the relation between sugar beet yields and air capacity. The response curve shows that below 8 or 10% air capacity yields are reduced. The authors comment that, "The results from the. . .experiments in 1940 were not as outstanding as in 1939 because of the small amount of rain that fell

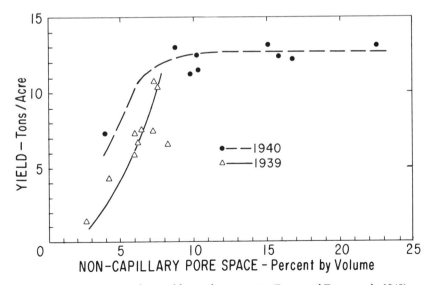

Fig. 1. Relation of sugar beet yields to soil air capacity (Baver and Farnsworth, 1940).

in June and July. . .1940 as compared with. . .1939." They recognized that the air porosity effects changed with weather conditions. Table 1 also shows that the air capacity changed during the season due to the subsidence of the soil and cannot be considered a constant. The actual air porosity or air fraction is also continuously changing with rainfall and dry periods even if the matrix is stable. During periods of intense rainfall the volume fraction filled with gas is greatly reduced.

Russell (1949) described an air pycnometer which could be used in the field to measure the volume of air-filled pores under field conditions. He also demonstrated its use on tillage plots.

Ehlers (1973) determined complete pore size distributions on tilled and untilled field plots. The greatest changes due to tillage were in the large pores ($> 30 \mu$) which are not related to air capacity because of their large size and physical instability.

Reviewers Grable (1966), Stolzy (1974), and Wesseling and Van Wijk (1957) after discussing the available data suggest that the minimum limit of air capacity in soils for the growth of common crops is about 10%. Grable (1968) after considering Van Duins data concludes that, "No single value of porosity can be considered optimum, or even minimum, for all situations."

The effect of tillage could be evaluated by following air porosity changes with time by making volumetric soil moisture measurements and calculating air porosity. However, these studies would have dubious results because the soil matrix is subject to change after tillage and the limits of porosity for normal plant growth is questionable.

Air Composition

The aeration intensity parameter or soil air composition and its relation to plant growth has been reviewed by Russell (1952), Currie (1962), Grable, 1966, 1968), Greenwood (1969), Taylor et al. (1972), Stolzy (1974), and Armstrong (1979).

The measurement of soil gas composition is actually a measurement of gaseous disequilibrium between the soil and atmospheric air (Tiedje et al., 1973). The biological activities in the soil are both a sink for oxygen and a source of CO_2. Were it not for soil impedance to diffusive mixing, the disequilibrium would be much less, and the composition of soil air would be similar to that of atmospheric air.

Tillage can drastically alter the soil matrix, changing the configuration of pores. This interacts with the soil water regime to modify the amount of gas-filled pores as well as the tortuosity and continuity of pores. Currie (1962) concludes that "soil cultivation can help aeration and hence fertility. By increasing the non-capillary pore volume, they improve air conditions within the primary aeration path of the soil profile. By breaking up the matrix into smaller units, they can influence the air composition within these units and promote more active root growth. Cultivation cannot, except under adverse conditions, affect the structure within these units, but by improving aeration they can improve permeability to roots and so lead to the necessary changes."

The relation of soil air composition to tillage and matrix changes is not predictable because of the biological source-sink activity in the soil. This activity is not constant. Its variability may or may not be related to the tillage operation. Biologically, the measurement of oxygen concentration is used because "oxygen is an invariant requirement for root growth" (Armstrong, 1979). However, the roots are within the water phase which can be a formidable barrier to oxygen transfer (Wiegand and Lemon, 1958).

The limits of root growth responses to variations in gaseous oxygen concentration has been reviewed by Armstrong (1979) and Taylor et al. (1972), but the data is rather limited. Williamson and Splinter (1968), who grew tobacco (*Nicotiana tabacum* L.) in a mist chamber, observed no reduction in root or shoot growth at 2.5% oxygen. Oxygen at 1% with 20% CO_2 for 1 or 2 days caused death of root tips, chlorosis of lower leaves, and growth reduction during treatment. Treatments with 0% oxygen and 21% CO_2 for 1 day killed the plants. Huck (1970) flushed the gases in the soil environment in which soybeans (*Glycine max* L.) and cotton (*Gossypium hirsutum* L.) were growing. Roots grew normally at oxygen concentrations above 10%. Two to five percent oxygen reduced tap root elongation, but elongation rates recovered when normal air was retained. Anoxia caused the death of root tips in 3 hours for cotton and 5 hours for soybeans. Campbell and Phene (1977) grew millet (*Pennisetum americanum* (L.) Leeke] on a tillage- irrigation field experiment and found normal growth above 15% oxygen and yield response within the range of 2 to 15%. Meek et al. (1980), found that summer flooding of alfalfa (*Medicago sativa* L.), which decreased the oxygen concentration to less than 6%, caused substantial losses in alfalfa stands. They cite other data where alfalfa was not thought to be this susceptible. The data were taken by Meek during the heat of summer. Temperature stress both above and within the soil must be considered in this situation. A limit for oxygen concentration for normal plant growth is not possible because of the limited data available, the influence of other stresses and the importance of moisture film thickness in field soils.

The classic work of Russell and Appleyard (1915) followed the oxygen and CO_2 concentration changes in natural soils throughout the year. They found changes with season, soil, crop, and tillage. Boynton's (1941) pioneering work on soil atmosphere in New York state demonstrated the influence of rainfall.

Since these studies, there has been improvement in methods of measurement and sampling (Letey, 1965), Raney (1949) used the paramagnetic method to measure oxygen concentrations in samples taken from gas wells to study oxygen changes in tillage experiments. Since then, the gas chromatograph has become available (Tackett, 1968). Much smaller samples are needed which can be withdrawn from small diffusion wells or directly from the soil with hypodermic needles (Lai et al., 1976). Membrane covered electrodes can be installed in the soil (Willey and Tanner, 1963; Patrick, 1977) and used to follow oxygen concentrations in situ.

Raney (1949) measured oxygen concentrations on a cropping-system tillage experiment 6 weeks after tillage. Oxygen concentrations in the

Table 2. Summary of the effects of tillage and depth on some physical properties
of Ovid Silt Loam. †

Depth inches	Tillage implement	Partial pressure oxygen in soil air	Diffusion rate
4	Subsoil plow	141	0.546
4	Regular plow	141	0.639
4	Disc	138	0.377
4	Rotary tiller	134	0.356
Average		139	0.479
L.S.D.		4.0	0.094
8	Subsoil plow	137	0.428
8	Regular plow	138	0.417
8	Disc	129	0.279
8	Rotary tiller	121	0.209
Average		131	0.333
L.S.D.		4.5	0.028

† From Raney (1949).

Table 3. Effect of amount of water and rainfall on soil oxygen and matric potential. †

Treatment symbol	Mean soil oxygen	Mean soil matric potential	Rainfall plus irrigation
	%	mb ± S	cm
Very wet	5.5 ± 3.4	−45 ± 14	31
Wet	9.8 ± 2.6	−57 ± 8	16
Mod. wet	13.2 ± 4.2	−87 ± 27	10
Mod. dry	18.7 ± 0.5	−296 ± 100	8

† From Campbell and Phene (1977).

plow zone were lower on the rototilled plots than on the plowed plots (Table 2). Burnett and Tackett (1968), measured oxygen concentration profiles to a depth of 6 feet on soils that had been mechanically mixed to a 4 foot depth and found the oxygen to be more uniform in modified profiles. Richter (1974) measured CO_2 concentrations in soils that were nontill planted and rototilled. He found higher CO_2 levels in the non-tilled which he attributed to enhanced biological activity.

Campbell and Phene (1977) performed a tillage experiment on a problem soil that had a thick compact layer just below the plow layer. Conventional plowing 17 cm deep and chiseling to 38 cm deep were compared under irrigated or non-irrigated conditions with millet as the crop harvested as forage. Oxygen concentrations were followed with membrane electrodes. Table 3 is their mean data for moisture regimes and shows the decreases in average oxygen concentration as the moisture regime becomes wetter. Figure 2 shows how the concentration of oxygen changed with water additions throughout the season at two depths and illustrates the cycling of the oxygen concentration parameter. Figure 3 shows the relation between relative yield and average oxygen concentration. It shows that, below 15% oxygen, millet yields decreased with decreasing oxygen concentrations. Above 15%, the yields and concentrations were similar for the tillage treatments.

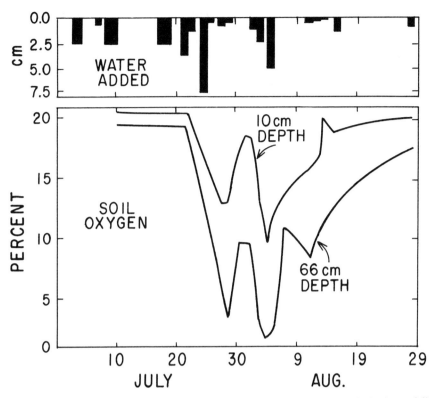

Fig. 2. Oxygen content in an irrigated field soil followed by a moderately high rainfall (Campbell and Phene, 1977).

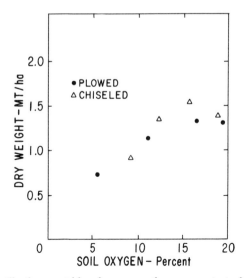

Fig. 3. Relative millet forage yield and average soil oxygen content of plowed and chiseled plots (Campbell and Phene, 1977).

Dowdell et al. (1979) compared plowing and direct drilling and followed the soil oxygen concentrations at three depths for three growing seasons. The mean oxygen concentrations varied with the seasonal rainfall. Direct drilling resulted in higher oxygen concentrations than plowing during the wetter winters, averaging 10.2 and 7.2%, respectively. The higher oxygen concentrations found in direct drilling were possibly due to a system of large pores and channels which developed in the direct drilled plots but which were destroyed by annual plowing.

Diffusion in the Gas Phase

Diffusion is the process by which gases are exchanged in the soil. Carbon dioxide from biological respiration moves from the living tissues via the soil air space to the atmospheric sink, while oxygen from the atmosphere moves in. The classic work of Buckingham (1904) considered the measurement of a diffusion coefficient in soil materials and its relation to soil porosity. Since then a great deal of attention has been given to similar studies. Reviews which consider diffusion include those of Currie (1962), Letey (1965), Grable (1968), Stolzy (1974), and Currie (1979).

Because tillage alters the soil matrix, which in turn influences the gaseous diffusion coefficients, diffusion in the gas phase seems to be a reasonable parameter to be measured. The problem lies in relating this parameter to crop growth. Taylor[4] has suggested that an approximate value of $D/D_o = 0.111$ for the oxygen diffusion rate in the gaseous phase may be a critical point where the plants begin to suffer from lack of aeration. Grable (1968) did an excellent analysis of the problem. After developing a possible critical value for the relative diffusion coefficient of about 0.02 for maximum root elongation he concludes with this statement: "The parameter D/D_o pertains only to gaseous diffusion through bulk soil. Plant growth is also dependent on gaseous diffusion through water films surrounding roots and microorganisms. Therefore, a single critical value of D/D_o for plant growth does not exist." The paper of Wiegand and Lemon (1962) emphasized the importance of the water film thickness in controlling oxygen availability to the root.

Raney (1949) developed a field method for measuring D/D_o and used it to evaluate a tillage experiment. Table 2 from Raney (1949) is given as an illustration. At a depth of 4 inches, diffusion rates, D/D_o, in a soil loosened by plowing is one-half to two-thirds of that in the atmosphere. The soil, disked and rotary tilled, has a D/D_o of aobut one-third. At the 8 inch depth at D/D_os are less but relatively the same. These measurements are consistent with what is known about soils after these tillage operations. The plow loosens, the disc compacts, and the rototiller breaks the soil structure so fine that it quickly settles.

Richter (1974) and Richter and Grossgebauer (1978) measured D/D_o on tillage plots on clay soils in forest nurseries and found the apparent diffusion to be higher in the untilled plots when compared to the tilled plots. This experiment demonstrated the deleterious effects of repeatedly tilling these soils.

---

[4]Taylor, S. A. Soil air-plant growth relations with emphasis on means of characterizing soil aeration. Doctor's Dissertation. Cornell University (1949).

Currie's work (1963) based on the measurement of diffusion coefficients has shown that most structured soils are bimodal with regard to this parameter. It is the crumbs which are more or less stable and have their own particular relative diffusion coefficients. Tillage affects the intercrumb structure and this arrangement alters the other part of the apparent diffusion coefficient both of which are measured by the D/D_0 term in the dry state.

Lai et al. (1976) developed a micro in situ soil gas diffusion measurement for a specific purpose which could be useful in some situations. It is probably not adapted for use in loose tilled soils because of their heterogeneity.

Diffusion within the Soil Moisture Films

"Plant roots require an adequate supply of oxygen for respiration in order to facilitate the maximum absorption of water and nutrients. If the rate of supply of oxygen to the root surfaces is limiting, plant growth will be reduced. . . .As the active root surfaces are covered with water films, this interface is the cell-wall liquid-phase boundary of the root and the water film. Movement of oxygen from the atmosphere to the actively respiring cells of the plant root involves not only diffusion through the gaseous phase of the soil but also movement through the gas liquid-phase boundary and liquid phase of the water film" (Lemon and Erickson, 1952).

"In the water-film cell-wall portion of the oxygen chain, movement is probably by diffusion only. Under normal conditions the direction of oxygen diffusion is down a concentration gradient from the soil atmosphere into and through the water films of the soil to cell walls of the root. This concentration gradient may at times, if not always, be very marked. This suggests then that some method of measuring oxygen diffusion, down a considerable concentration gradient, through the liquid phase to a reducing surface similar to that of a plant root, would be of fundamental importance" (Lemon and Erickson, 1952).

This premise has inspired experimental and theoretical studies to better elucidate the relations between the root, the soil moisture films, and the soil atmosphere. Wiegand and Lemon (1958) in a theoretical and experimental study demonstrated the importance of the "apparent diffusion path length" in the liquid phase about plant roots in limiting normal root respiration. They also concluded that certain root surfaces in a Miller clay would be suboptimal when the soil was at field capacity. Other theoretical papers of the plant soil relation include Lemon (1962), Lemon and Wiegand (1962), Luxmore et al. (1970), and Luxmore and Stolzy (1972). Reviews on the subject have been made by Currie (1962), Vilain (1963), Letey et al. (1963), Erickson (1965), Grable (1966, 1968), Stolzy (1974), and Armstrong (1979). The general conclusion is that the ultimate soil aeration parameter to describe soil-air plant relations is the measure of the rate at which oxygen diffuses to the actively respiring plant root surface within the water phase. The water phase is a formidable barrier because the oxygen diffusion coefficient in water is 0.0001 and the solubility is 0.03 of that in air. In order to maintain sufficient oxygen at the root surface for normal respiration, a critical rate of resupply is necessary. This rate of supply is called the critical oxygen diffusion rate, which can also be expressed as a critical moisture film thickness (Wiegand and Lemon, 1958; Luxmoore et al., 1970).

Lemon and Erickson (1952) also demonstrated a platinum microelectrode method to simulate actively respiring plant roots within the water films. The method, based on polarography, measures the rate at which oxygen is reduced at the electrode surface which is an absolute sink for oxygen. This rate, referred to as the oxygen diffusion rate (ODR), is proportional to the current drawn by the electrode, and hence quantifiable.

Lemon and Erickson (1955), Letey et al. (1963), Stolzy and Letey (1964), Van Doren and Erickson (1966), and Fluhler et al. (1976) have described the method, its advantages, disadvantages, and limitations. McIntyre (1970) made a severe critique of the method, but the previous reviews and subsequent reviews by Stolzy (1974), Smith (1977), and Armstrong (1979) seem agreed that over the moisture range in which ODR is critical for normal plant activity the method is probably functioning properly. Phene et al. (1976) published a method for automating ODR measurements which would allow for frequent measurements during the dynamic drainage periods when oxygen deficiency is most likely to occur.

The relation of ODR to plant growth has been reviewed by Stolzy and Letey (1964), Erickson (1965), Stolzy (1974), and others. The data indicate that at ODRs below 0.2 μg cm^{-2} min^{-1} plant roots will not grow, plants are severely stressed and may die; between 0.2 and 0.4 μg cm^{-2} min^{-1} the plants are retarded and above 0.4 μg cm^{-2} min^{-1} plants grow normally.

Erickson and Van Doren (1960) showed that short-term oxygen deficiency of 1 day could reduce the ultimate growth by 30 or 50% for peas (*Pisum sativum* L.) and tomatoes (*Lycopersicon esculentum* Mill.), respectively. Other studies by Fulton and Erickson (1964) and Bolton and Erickson (1970) demonstrated that photosynthetic and evapotranspiration stress concurrent with the oxygen stress will further increase damage to the crop. This is similar to field data reported by Letey et al. (1963), Meek et al. (1978), and Meek et al. (1980).

The platinum microelectrode method can be used to measure the effect of tillage on soil aeration in terms that are related to crop growth. This has been done by Erickson and Van Doren (1960) and Phene et al. (1976) with different objectives and was proposed by Erickson (1965) to evaluate drainage systems. Oxygen diffusion rate measurements were made during each rain or irrigation period. If the measured ODRs were below the critical value of 0.35 μg cm^{-2} min^{-1}, readings would continue until the soil had adequate aeration. The accumulated deficiency days or better a combination of deficiency duration and intensity could be used as a measure of stress. Hiler et al. (1971) suggested an oxygen stress index comparable to the water stress day index. A further improvement in predicting effects of oxygen deficiency stress would be to include the type of crop, its stage of development, and the atmospheric stress, temperature, and evapotranspirative demand, during the oxygen stress period.

Table 4. Oxygen diffusion rates from a tillage experiment on Charity clay
after a 7-cm rain on 27 July 1980.

Date	Deep chiseled	Plowed	Compact
	$\mu g\ cm^{-2}\ min^{-1}$		
July			
28	0.32	0.29	0.18
29	0.34	0.30	0.15
30	0.40	0.31	0.22
31	--	--	--
August			
1	0.43	0.35	0.31
2	--	--	--
3	--	--	--
4	0.46	0.38	0.37

An example of this method applied to a tillage experiment is given in
Table 4 (A. E. Erickson, unpublished). This soil is a Charity clay (Aeric
Haplaquept) which has poor structure and is so dense that little rooting
occurs below the plow zone except in the shrinkage cracks that occur late
in the season. The tillage treatments were (1) deep tilled (chiseled in the
fall 50 cm deep and fall plowed), (2) fall plowed (fall plowed), and (3)
compacted (fall plowed, compacted with tractor tire in the spring). All
plots were no-till planted in the spring. The early season was dry but
turned wet. The 7-cm rainfall on 27 July 1980 caused flooding of all treatments. Periods of oxygen deficiency were 2 days, 5 days, and between 6
and 8 days for the deep chiseled, fall plowed, and compact treatments,
respectively. The readings should have been more frequent. The yield of
beans (*Phaseolus vulgaris* L.) were in the same order but were also influenced by drought stress early in the season.

Biochemical Methods

Biochemical methods for measuring oxygen stress in plants should be
mentioned. The ethanol concentration in the xylem exudate of stressed
plants (Fulton and Erickson, 1964; Bolton and Erickson, 1970) uses the
plant as an integrator of aeration stress. Ethylene in the poorly aerated
soil atmosphere is discussed by Sheard (1976) and Smith (1977). It ac-
cumulates in anoxic soils and is known to affect root growth and seed
germination.

CONCLUSIONS

Russell's (1952) conclusion that "much of the information relating
soil aeration to crop growth is based on inference rather than quantitative
data", need no longer impair tillage research. The platinum
microelectrode method can be used to describe the soil root environment
in terms that affect plant growth.

Tillage experiments that caused Hawkins (1962) to lament, "field experiments, in which the effects of cultivation are expressed only in terms

of crop yields, provide little information of value to guide the design of cultivation equipment or to answer general cultivation problems," should no longer be acceptable.

LITERATURE CITED

1. Armstrong, W. 1979. Aeration in higher plants. p. 225–332. *In* Advances in botanical research. Vol. 7. Academic Press, London.

2. Baeumer, K., and W. A. P. Bakermans. 1973. Zero tillage. Adv. Agron. 25:77–123.

3. Baver, L. D., and R. B. Farnsworth. 1940. Soil structure effects on the growth of sugar beets. Soil Sci. Soc. Am. Proc. 5:45–48.

4. Bolton, E. F., and A. E. Erickson. 1970. Ethanol concentration in tomato plants during soil flooding. Agron. J. 62:220–224.

5. Boynton, D. 1941. Soils in relation to fruit growing in New York. Part XV. Seasonal and soil influences on oxygen and carbon dioxide levels of New York orchard soils. p. 1–43. Cornell University Agric. Exp. Stn. Bull. 763.

6. Buckingham, E. 1904. Contribution to our knowledge of aeration in soils. USDA Bur. of Soils, Bull. 25.

7. Burnett, E., and V. L. Hauser. 1968. Deep tillage and soil-plant-water relations. p. 47–52. *In* Tillage for Greater Crop Production. ASAE, St. Joseph, Mich.

8. ————, and J. L. Tackett. 1968. Effect of soil profile modification on plant root development. p. 329–337. Proc. 9th Int. Soils Congr. Adelaide, Australia. Vol. 7.

9. Campbell, R. B., and C. J. Phene. 1977. Tillage, matric potential, oxygen and millet yield relations in a layered soil. Trans. ASAE 20:271–275.

10. Cannell, R. Q. 1977. Soil aeration and composition in relation to root growth and soil management. Appl. Biol. 2:1–86.

11. Currie, J. 1962. The importance of aeration in providing the right conditions for plant growth. J. Sci. Food Agric. 13:380–385.

12. ————. 1979. Rothamsted studies of soil structure IV. Porosity, gas diffusion and pore complexity in dry soil crumbs. J. Soil Sci. 30:441–452.

13. Dowdell, R. J., R. Cress, J. R. Burford, and R. Q. Cannell. 1979. Oxygen concentrations in a clay soil after ploughing or direct drilling. J. Soil Sci. 30:230–245.

14. Ehlers, W. 1973. Total porosity and pore size distribution in untilled and tilled loess soils. Z. Pflanzenernaehr. Bodenkd. 134:193–207.

15. Erickson, A. E. 1965. Short-term oxygen deficiencies and plant response. p. 11–12, 23. *In* Drainage for efficient crop production conference. ASAE, St. Joseph, Mich.

16. ————, and D. M. Van Doren. 1960. The relation of plant growth and yield to soil oxygen availability. 7th Int. Congr. Soil Sci. III:428–434.

17. Fulton, J. M., and A. E. Erickson. 1964. Relation between soil aeration and ethyl alcohol accumulation in xylem exudate of tomatoes. SSSA Proc. 28:610–614.

18. Fluhler, H., M. S. Ardakani, T. E. Szuszkiewicz, and L. H. Stolzy. 1976. Field-measured nitrous oxide concentrations, redox poentials, oxygen diffusion rates and oxygen partial pressures in relation to denitrification. Soil Sci. 122:107–114.

19. Grable, A. R. 1966. Soil aeration and plant growth. Adv. Agron. 18:57–106.

20. ————. 1968. Effects of tillage on soil aeration. p. 44–55. *In* Tillage for greater crop production. ASAE Mon., St. Joseph, Mich.

21. Greenwood, D. J. 1969. Effect of oxygen distribution in the soil on plant growth. p. 202–223. *In* W. J. Whittington (ed.) Root growth. Plenum Press, New York.

22. Hawkins, J. C. 1962. The effects of cultivation on aeration, drainage, and other soil factors important in plant growth. J. Sci. Food Agric. 13:386–391.

23. Hiler, E. A., R. N. Clark, and L. J. Glass. 1971. Effects of water table height on soil aeration and crop response. ASAE Trans. 13:879–882.

24. Huck, M. G. 1970. Variations in tap root elongation rate as influenced by composition of soil air. Agron. J. 62:815–818.

25. Lai, S. H., J. M. Tiedje, and A. E. Erickson. 1976. In situ measurement of gas diffusion coefficient in soils. SSSA J. 40:3–6.

26. Lemon, E. R. 1962. Soil aeration and plant root relations. I. Theory. Agron. J. 54:167–170.

27. ————, and A. E. Erickson. 1952. The measurement of oxygen diffusion in the soil with a platinum microelectrode. Soil Sci. Soc. Proc. 16:160–163.

28. ————, and ————. 1955. Principle of the platinum microelectrode as a method of characterizing soil aeration. Soil Sci. 79:383–392.

29. ————, and C. L. Weigand. 1962. Soil aeration and plant root relations. II Root respiration. Agron. J. 54:171–175.

30. Letey, J. 1965. Measuring aeration. p. 6–10. In Drainage for efficient crop production. ASAE, St. Joseph, Mich.

31. ————, and L. H. Stolzy. 1964. Measurement of oxygen diffusion rates with the platinum microelectrode. Hilgardia 35:545–575.

32. ————, ————, N. Valaras, and T. E. Szuszkiewicz. 1963. Low soil oxygen most damaging to plants during hot weather. Calif. Agric. 17:15.

33. Luxmoore, R. J., and L. H. Stolzy. 1972. Oxygen diffusion in the soil-plant system VI. A synopsis with commentary. Agron. J. 64:725–729.

34. ————, ————, and J. Letey. 1970. Oxygen diffusion in the soil-plant system I. A Model. Agron. J. 62:317–322.

35. Meek, B. D., T. J. Donovan, and L. E. Graham. 1980. Summertime flooding effects on alfalfa mortality, soil oxygen concentration and matric potential in a silty clay loam soil. Soil Sci. Soc. Am. J. 44:433–435.

36. ————, and L. H. Stolzy. 1978. Short term flooding. p. 351–375. In Donald Hook (ed.) Plant life in anoerobic environments. Ann Arbor Sci., Ann Arbor, Mich.

37. McIntyre, D. S. 1970. The platinum microelectrode method for soil aeration measurements. Adv. Agron. 22:235–283.

38. Patrick, W. H. 1977. Oxygen content of soil by a field method. SSSA J. 41:651–652.

39. Pearson, R. W. 1965. Soil environment and root development. p. 95–126. In Plant environment and efficient water use. Am. Soc. of Agron., Madison, Wis.

40. Phene, C. J., R. B. Campbell, and C. W. Doty. 1976. Characterization of soil aeration in situ with automated oxygen diffusion measurements. S.S. 122:271–281.

41. Raney, W. A. 1949. Field measurement of oxygen diffusion through soil. SSSA Proc. 14:61–65.

42. Richter, J. 1974. A comparative study of soil gas regimes in a soil tillage experiment with different soils. I. Field measurements. Z. Pflanzenernaehr. Bodenkd. 137:135–147.

43. ————, and A. Grossgebauer. 1978. The soil air regime in a soil tillage experiment. Part 2. Gas diffusion coefficient as a measure of soil structure. Z. Pflanzenernaehr. Bodenkd. 141:181–202.

44. Russell, E. J., and A. Appleyard. 1915. The atmosphere of the soil, its composition and causes of variation. J. Agric. Sci. 7:1–48.

45. Russell, M. B. 1949. A simplified air-picnometer for field use. SSSA Proc. 14:73–76.

46. ————. 1952. Soil aeration and plant growth. p. 253–301. In Soil physical conditions and plant growth. Academic Press, Inc., N.Y.

47. Sheard, R. W., and A. J. Leyshon. 1976. Short-term flooding of soil: its effect on the composition of gas and water phase of soil and on phosphorus uptake of corn. Can. J. soil Sci. 56:9–20.

48. Smith, K. A. 1977. Soil aeration. Soil Sci. 123:284–291.

49. ————, and R. J. Dowdell. 1974. Field studies of the soil atmosphere I. Relationship between ethylene, oxygen, soil moisture content and temperature. J. Soil Sci. 25:219–230.

50. Stolzy, L. H. 1974. Soil atmosphere. p. 335–361. In E. W. Carson (ed.) The plant root and its environment. University Press of Virginia, Charlottesville.

51. ————, and J. Letey. 1964. Characterizing soil oxygen conditions with a platinum microelectrode. Adv. Agron. 16:249–279.

52. Tackett, J. L. 1968. Theory and application of gas chromotography in soil aeration research. SSSA Proc. 32:346–350.

53. Taylor, H. M., M. G. Huck, and B. Klepper. 1972. Root development in relation to soil physical conditions. p. 57–77. *In* D. Hillel (ed.) Optimizing the soil physical environment toward greater crop yields. Academic Press, N.Y.

54. Tiedje, J. M., S. H. Lai, and A. E. Erickson. 1973. Martian life detection based on in situ gas analysis. Bull. Ecol. Res. Comm. Stockholm. 17:489–497.

55. Van Bavel, C. H. M. 1954. Simple diffusion well for measuring soil specific impedance and soil air composition. SSSA Proc. 18:229–234.

56. Van Doren, D. M., and A. E. Erickson. 1966. Factors affecting the platinum microelectrode method for measuring the rate of oxygen diffusion through the soil solution. Soil Sci. 102:23–28.

57. Vilain, M. 1963. L'aeration du sol. Ann. Agron. 14:967–998.

58. Weigand, C. L., and E. R. Lemon. 1958. A field study of some plant-soil relations in aeration. SSSA Proc. 22:216–221.

59. Wesseling, J., and W. R. VanWijk. 1957. Soil physical conditions in relation to drain depth. p. 461–504. *In* J. N. Luthin (ed.) Drainage of agricultural lands. Am. Soc. Agron., Madison, Wis.

60. Willey, C. R., and C. B. Tanner. 1963. Membrane-covered electrode for measurement of oxygen concentration in soil. SSSA Proc. 27:511–515.

61. Williamson, R. E., and W. E. Splinter. 1968. Effect of gaseous composition of root environment upon root development and growth of Nicatiana tobacum L. Agron. J. 60:365–368.

Chapter 7

Predicting Tillage Effects on Infiltration[1]

W. M. EDWARDS[2]

ABSTRACT

Although the hydraulic characteristics of the plow layer and deeper soil horizons are often used as a basis for calculating infiltration, surface characteristics usually govern the entry of water into the soil during high rainfall rates. The effects of surface characteristics and of surface connected noncapillary sized holes and cracks are usually not determined in traditional soil physics based infiltration models. Tillage operations influence both the surface (cover) and subsurface (soil) conditions. Under conventional tillage for row crops (unprotected surface), crusting is very important. In 203 May-June storms, runoff from crusted corn watersheds was nearly three times that from uncrusted fields. With continuous no-tillage management, plow layer porosity of a Typic Hapludult decreased from 50 to 40 %, but the preserved macrochannels in the soil and crop residue on the surface reduced runoff to 1/20 of that from a nearby conventionally tilled watershed. Tillage increased the immediately available pore space near the surface, but decreased the amount of infiltrating water moving deeper into the profile.

INTRODUCTION

The task of developing a comprehensive numerical model of the infiltration process on agricultural lands is formidable. Rainwater may infiltrate for a time, only to come back to the surface at a point downslope,

---

[1] Contribution of the North Appalachian Experimental Watershed, USDA-ARS, Coshocton, Ohio, in cooperation with the Ohio Agricultural Research and Development Center, Wooster, Ohio.
[2] Soil scientist, USDA.

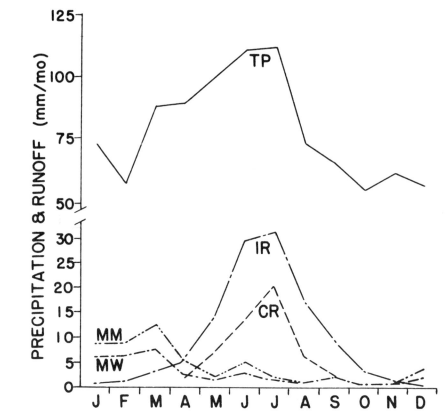

Fig. 1. Long-term averages at Coshocton, Ohio of total precipitation (TP), rainfall at intensities >25 mm/hour (IR), runoff from conventionally tilled corn on watershed 109 (CR), and runoff from well-drained (MW) and moderate drained (MM) meadow watersheds.

or it may run off the upslope area, then infiltrate in an area below. The time-dependent disposition of rainwater is very complex, being influenced by many factors discussed elsewhere in this symposium.

Tillage operations which accompany crop production have long been recognized as important determinants of the infiltration-runoff relationship. To help evaluate the effects of land management on hydrology, the North Appalachian Experimental Watershed was established in 1935 in the hills of east-central Ohio. For more than 40 years, rainfall and runoff have been measured from about 25 1 to 3-ha watersheds, giving about 150,000 measurements of runoff and infiltration from field-sized areas tilled with normal farming equipment.

The long-term hydrologic records from the Coshocton watersheds give direct insight into many of the effects of land management on infiltration. In addition, the runoff records show that factors which strongly affect infiltration under some conditions may have little effect under others. The objective of this paper is to define factors that influence the infiltration-runoff relationship under different conditions and to show

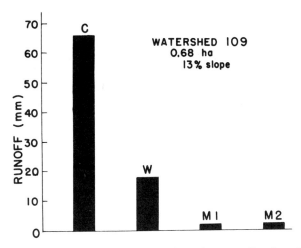

Fig. 2. Average annual runoff from watershed 109 farmed in corn (C), wheat (W), meadow (M1), meadow (M2) rotation, 1941 to 1969.

how these factors may be used to predict infiltration. Where rainfall-runoff records are used, we assumed that infiltration equals rainfall minus runoff.

Infiltration Influenced by Precipitation Pattern. The seasonal distribution of precipitation amount and intensity influences the kinds of crops that can be grown without irrigation in a given area and the kinds of management practices needed throughout the year to combat problems caused by runoff or excess soil water. Although rainfall directly influences both infiltration and runoff, the interactions among rainfall characteristics and other climatic factors, the soil surface, and soil conditions cause the correlation between rainfall and infiltration to vary widely.

The top two curves of Fig. 1 show long-term averages of monthly precipitation (mostly rainfall) at Coshocton and amount of rain falling at intensities > 25 mm/hour. Consistently, both intensity and amount are low in the fall and winter (dormant season) and reach annual maxima in June and July. Clean seedbeds for row crops are traditionally prepared in April and May, just before the intense rains. Growing vegetation or crop residues protect the surface during the rest of the year.

The effects of the vegetative cover on runoff and infiltration under this climatic regime are shown on a monthly basis in Fig. 1. Growing season runoff from the clean-tilled corn watershed (CR) is well correlated with the amount of rain that falls at intensities > 25 mm/hour (IR). However, that same relation does not hold in years that the surface is not tilled (Fig. 1). The watershed was farmed in a 4-year rotation (corn, wheat, meadow, meadow) for 29 years, being plowed, disked, harrowed, and cultivated for corn production every 4th year (Fig. 2). Runoff averaged 64 mm/year in the corn years, 19 mm/year in the wheat years, and less than 2 mm/year in the meadow years. In this rainfall regime, tillage and the associated differences in crop cover caused 30-fold differences in average annual runoff.

From an infiltration modeling standpoint, Fig. 1 shows two other important considerations. First, the low intensity rains of the fall and early winter at Coshocton cause very little runoff from the meadow fields (MM and MW). Storms of similar intensities in January, February, and March cause more runoff. The relationship of the rainfall pattern to the annual climatic cycle causes the difference. In the fall, the rain falls on soils recently dried by the strong evapotranspirational demand, which in this area is greater than summer rainfall. Thus, water storage capacity is available in these soils during the fall season. Winter temperatures reduce evaporation and transpiration, the soil profiles become wetter, and downward movement through flow-restricting subsurface soil layers limits infiltration. Also, the topsoil is often frozen in January and February.

Second, subsoil hydraulic characteristics affect infiltration when the profiles are nearly saturated, but they are not important at other times. The meadow runoff relations MM and MW of Fig. 1 are each based on 64 watershed years of record. Drainage characteristics of the soils of the MM watersheds are classified as moderate, while soils of the MW watersheds are well drained. During the first 4 months of the year, differences in hydraulic characteristics of subsoil horizons cause strong differences in runoff that are not apparent when the soil profiles are drier.

Several relationships important to modeling infiltration are indicated: (a) When the surface is bare and rainfall intensity varies, correlation between rainfall intensity and growing season runoff is good. Under similar rainfall patterns when the surface is well vegetated, that relationship does not hold; (b) Low intensity rains do not cause runoff from protected surfaces when storage capacity is available in the soil. As the profile becomes saturated, flow-restricting layers in the subsoil can limit infiltration, causing runoff from low intensity rainfall; and (c) Most important is the observation that the factors which influence infiltration and runoff vary within the year and infiltration models must be responsive to these cyclic changes.

Infiltration into Tilled Surfaces. From the standpoint of runoff quality, infiltration in the early part of the cropping season is most important. In many modern row-crop farming systems, herbicides are used to control weeds, replacing one or more mechanical cultivations which were needed for that job before adequate herbicides became available. In addition to controlling weeds, cultivation also broke up any surface crust that existed at the time that weed control was needed. With herbicidal weed control the crust is no longer broken periodically. The first crust that develops after planting persists throughout the growing season.

To evaluate the effect of crusting upon infiltration and runoff, we examined the May-June storms in corn years of eight rotation watersheds over a 24-year period. Physical and chemical characteristics of the soils in these watersheds are described elsewhere (8). During May and June, the clean-tilled soil is subject to crusting because the corn canopy is sparse and the rain intensity is high.

For this study, we arbitrarily defined uncrusted soils (U) as those receiving < 13 mm of rain between the last tillage or cultivation and the start of a storm. The crusted soils (C) were defined as those receiving 38

Table 1. Measured rainfall and runoff from a corn watershed in June 1960, and predicted runoff assuming no cultivation on 18 June.

Date	Rainfall	Runoff	
		Observed	Predicted
		mm	
June			
13	62.2	10.7	
14	14.7	5.0	
18	(cultivated)		
22	16.0	0.1	2.4
29	12.2	0.0	1.6

mm or more of rain since the preceding tillage operation. The data base contained 56 and 147 storms, respectively, that met these conditions for uncrusted and crusted antecedent conditions. Storms having > 13 but < 38 mm of antecedent rainfall since the last tillage were not included in this analysis.

Stepwise multiple regressions were run with runoff (RO) as the dependent variable and precipitation amount (P), antecedent topsoil moisture (M), and indexes of vegetative cover, rainfall intensity, and soil permeability as independent variables. The squares and interaction terms of the independent variables were also included. The best fit equations selected to predict storm runoff volume were:

$$RO_U = 6.60 - 0.223\,P + 0.00748\,P^2 - 0.287\,M + 0.00366\,M^2 \qquad [1]$$

$$R^2 = 0.74 \qquad \text{Standard deviation (SD)} = 1.32\,\text{mm}$$

$$RO_C = -0.254 + 0.130\,P + 0.00209\,P^2 \qquad [2]$$

$$R^2 = 0.77 \qquad SD = 2.18\,\text{mm}$$

In developing these equations, rainfall amount and intensity were equally important for predicting runoff or infiltration. But because neither added to the prediction capability when the other was included and because rainfall amount data are more readily available than are intensity data, the intensity index was not used in the equations.

To test the statistical significance of runoff from crusted vs. uncrusted conditions, the crusted soil equation was used to predict runoff for the 56 events when the soil was in fact uncrusted. The observed and predicted values were statistically different at the 99% level with total predicted runoff (162.4 mm) being almost four times that observed (42.7 mm).

Average observed runoff from these 203 runoff-producing storms in May and June at Coshocton was 2.0 mm. From the prediction equations, the average May-June storm falling on crusted soil would produce 2.2 mm of runoff, 15% of the 14.5 mm average rainfall per storm. The equations estimated no runoff from the same storm falling on uncrusted soil.

Table 1, including data from four storms in June 1960, illustrates one application of the simple model of Eq. [1] and [2] and shows the strong ef-

Table 2. Annual runoff from conventional and no-tillage corn watersheds.

| Year | Runoff | |
	Conv.†	No-till.‡
	mm	
1965	27	0
1966	27	0
1967	34	1
1968	34	0
1969	97	5
1970	14	0
1971	41	40
1972	55	0
1973–1978	not measured	
1979	77	4

† A different watershed each year, cropped in corn, wheat, meadow, meadow rotation.
‡ The same watershed each year cropped in continuous corn.

fect of cultivation upon runoff and infiltration. The first storm generated a crust and caused 10.7 mm of runoff. On the following day, because of the crust and the high antecedent soil water content, 14.7 mm of rain caused 5.0 mm of runoff.

Four days later the field was cultivated destroying the crust. On 22 June, a storm similar to the one on 14 June fell, but caused only 0.1 mm of runoff. Assuming that the crust had not been destroyed by cultivation, predicted runoff for that storm was 2.4 mm, over 20 times the observed amount. The difference between observed and estimated runoff for another storm on 29 June reinforces the concept.

Observed runoff from this watershed during the 17-day period (13 to 29 June 1960) totaled 15.8 mm with only 0.1 mm coming after the cultivation. It is estimated that runoff from the two latter events would have been 4.0 mm or 40 times more than observed, had the field not been cultivated on 18 June.

Equations [1] and [2] show that surface characteristics are important for predicting infiltration into the clean-tilled surfaces during the growing season. Data from these same watersheds were also examined to show how other parameters become more important for estimating infiltration during the dormant season.

Multiple regression analysis of 184 storms in January and February, when the watersheds had a cover of 10 to 20-cm-high winter wheat, gave a prediction equation ($R^2 = 0.82$) to which the soils index (9) contributed very significantly. The rainfall intensity variable, which was important for predicting May-June runoff, was not important for the January-February predictions. The soils index is based on soil drainage and permeability classes, both of which reflect subsurface characteristics (9).

Infiltration into Non-tilled Surfaces. It has been well established that the development of a crust on a bare soil surface decreases infiltration and that an infiltration model must recognize crusted vs. noncrusted conditions. Many of our tillage systems have been designed to help prevent the formation of soil crusts. The widening application of no-tillage as a practice, especially for production of row crops, has great potential for increasing infiltration under many conditions.

Table 2 shows how no-tillage as a management practice for corn production reduced runoff as compared with conventional tillage during a 15-year period at Coshocton. One watershed that was in continuous no-tillage corn every year of the study is compared with six conventionally tilled watersheds farmed in 4-year rotation (8). In every year, except 1971, tillage treatment strongly affected runoff. However, in 1971, when 80% of the total runoff from the no-tillage watershed occurred, tillage had no effect upon runoff. All of the 1971 runoff from both watersheds came during a few days in February as a result of moderate rainfall on snow-covered frozen soils.

The 1971 data describe an observed condition under which no-tillage as a management practice failed to increase infiltration at Coshocton. Elsewhere, other soils, weather patterns, and mulching practices produced various results. Average annual runoff from a claypan soil in central Missouri was greater with no-tillage management than with conventional tillage (11).

The no-tillage system produces several definable factors or conditions that may be used to estimate infiltration or runoff. The above ground effects of the cover are numerous. Among other things, the cover slows runoff, decreases the impact of the falling raindrops on the surface, and maintains continuity of open channels in the soil to the surface.

Below the surface, bulk density, a measure of total porosity, is one soil physical property that may influence infiltration. The no-tillage bulk density data of Fig. 3 were taken during the 15th and 16th years of continuous no-tillage corn production in the watershed from which the no-tillage runoff data of Table 2 were taken. Density in this watershed in 1966 consistently averaged <1.40, but with continued no-tillage it has gradually increased to 1.60.

Although the bulk density in 1978 and 1979 under no-tillage was consistently higher than that in the conventionally tilled watershed, runoff was nearly 20 times greater from the conventionally tilled watershed (Table 2). Infiltration was greater into the no-tillage topsoil, which had a total porosity of 40% at the 5 to 15-cm depth, than into the conventional topsoil, which had 50% total porosity, showing that bulk density is not always a good indicator of infiltration potential.

Figure 3 also shows that wheel traffic influenced bulk density under conventional tillage, but not under the long-term no-tillage treatment. If a relation between bulk density and infiltration can be shown for a specific soil, then size and weight of tractors and implements, which influence the degree of compaction and the proportion of packed and unpacked rows, become relevant. Topsoil strength in the no-tillage watershed, which was last plowed in 1960, is great enough that further compaction from current machine traffic can not be detected at the 5 to 15-cm depth. Therefore, an established relationship between bulk density and predicted infiltration for these soils is not subject to as much variation under no-tillage management as under conventional tillage.

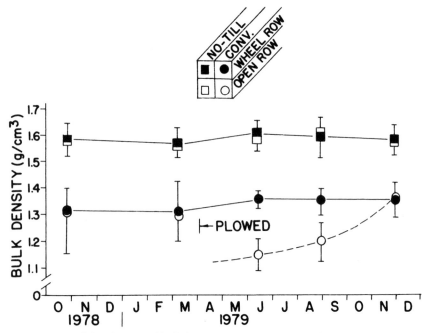

Fig. 3. Seasonal distributions of bulk density (mean and range, 40 determinations per date) measured at the 5 to 15 cm depth under long-term no-tillage and conventional corn.

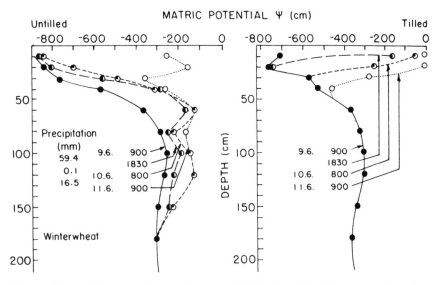

Fig. 4. Water infiltration and redistribution in tilled and untilled silt loam as indicated by changes in matric potential. From Ehlers and Baeumer (6).

Infiltration Influenced by Macropores. High infiltration rates into non-tilled surfaces have been related to a continuous system of macropores open to the surface (1, 12, 13). Ehlers (5) showed that the macropores in a loessial soil of West Germany were made by worms and that conventional tillage destroyed these holes in the plow layer, greatly reducing depth of infiltration. Figure 4, from Ehlers and Baeumer (6), clearly shows that infiltrating water moved deeper into the untilled soil soon after rainfall than it did in an adjacent conventionally tilled plot. Using dye as a tracer, Ehlers (5) showed that the deep, rapid penetration took place entirely in the worm holes that were open to the soil surface. Under forest conditions, similar flow in root channels has been reported (3).

These data show the need for including macropore components in infiltration models when the hydraulic characteristics of the soil matrix are only a part of the infiltration process. Traditional laboratory determinations of the water content-tension-conductivity relationship define the diffusion flow potential of the soil matrix, but they do not evaluate the effects of large pores upon infiltration.

Edwards et al. (4) modeled the field situation described by Ehlers (5) and calculated infiltration into soil columns having a central, round, vertical macropore. Their numerical solution used Darcy's Law and a radial coordinate system to describe vertical infiltration through the surface into the soil matrix. When the rainfall rate was greater than the infiltration rate, excess surface water ran into the hole and infiltrated radially away from the hole according to the water content-tension-conductivity relations of the relatively drier soil below the saturated surface. In this model, hole depth, diameter, and hole spacing influenced modeled infiltration and waer distribution in the soil.

Hoogmoed and Bouma (7) used similar techniques to model infiltration into a soil having cracks instead of round holes for macropores. Like Ehlers (5), they used a blue stain to indicate the location and magnitude of the large pore contribution to the infiltration process. Water movement was simulated as vertical infiltration into the surface soil, downward flow in the cracks, and horizontal infiltration away from the cracks below the surface. The volume of soil cracks contributing to total infiltration increased with increasing excess surface water.

Scotter (10) included both types of preferential ways, vertical cylindrical channels and planar cracks, in a model and examined the effectiveness of different sizes of each. His model predicted that significant preferential movement of both nonadsorbed and strongly adsorbed solutes could occur in continuous vertical channels at least 0.2 mm in diameter or in cracks at least 0.1 mm wide.

The total effects of surface-connected macropores on infiltration are strongly influenced by the hydraulic characteristics of the micropore system. This is because the relative volume of macropores is small, and once they are filled, further infiltration into the macropores depends upon the rate of movement into the surrounding micropore system (2).

In both the hole and the crack models, the numerical procedures seem adequate to describe water movement into and within a nonchanging soil matrix containing well-defined macropores. In both cases, however, a potential limitation to widespread application is the inadequate definition of the macropore system, including spacial variation, and the effects that tillage have on it.

SUMMARY

Examples of different factors affecting infiltration and runoff have been presented and discussed. The precipitation pattern interacts seasonally with surface conditions and soil characteristics, causing strong shifts in relative importance of factors influencing infiltration. Differences in permeability of subsoil layers may be important at the end of the dormant season when the overlying layers are nearly saturated, but they rarely influence infiltration during summer and fall rains when surface layers have water storage capacity.

Surface factors, such as soil crusts, mulch, or growing vegetation, are important variables in the summer months when rainfall intensity may be high. We can manage these surface and topsoil characteristics with soil tillage operations. For some crop and soil combinations, we can effectively manage infiltration by omitting the tillage operations that destroy the continuity of surface-connected macropores.

Infiltration can be predicted with simple or sophisticated models for specific or general application with the success of predictions depending heavily on the accurate determination of which factors are important under different conditions.

REFERENCES

1. Anderson, J. L., and J. Bouma. 1973. Relationships between saturated conductivity and morphometric data of an argillic horizon. Soil Sci. Soc. Am. Proc. 37:408–413.
2. Bevin, K., and P. Germann. 1981. Water flow in soil macropores: 2. A combined flow model. J. Soil Sci. 32:15–39.
3. de Vries, J., and T. L. Chow. 1978. Hydrologic behavior of a forested mountain soil. Water Resour. Res. 14:935–942.
4. Edwards, W. M., R. R. van der Ploeg, and W. Ehlers. 1979. A numerical study of the effects of noncapillary-sized pores upon infiltration. Soil Sci. Soc. Am. J. 43:851–856.
5. Ehlers, W. 1975. Observations on earthworm channels and infiltration on tilled and untilled loess soil. Soil Sci. 119:242–249.
6. ————, and K. Baeumer. 1974. Soil moisture regime of loessial soils in Western Germany as affected by zero-tillage methods. Pak. J. Sci. Ind. Res. 17:32–39.
7. Hoogmoed, W. B., and J. Bouma. 1980. A simulation model for predicting infiltration into cracked clay soil. Soil Sci. Soc. Am. J. 44:458–461.
8. Kelley, G. E., W. M. Edwards, L. L. Harrold, and J. L. McGuinness. 1975. Soils of the North Appalachian Experimental Watershed. USDA Misc. Pub. no. 1296, U.S. Government Printing Office, Washington, DC. 145 p.
9. McGuinness, J. L., and W. M. Edwards. 1975. A watershed-soils index of runoff potential. J. Soil Water Conserv. 30:184–186.
10. Scotter, D. R. 1978. Preferential solute movement through larger soil voids. I. Some computations using simple theory. Aust. J. Soil Res. 16:257–267.
11. Smith, G. E., R. Blanchar, and R. E. Burwell. 1979. Fertilizers and pesticides in runoff and sediment from claypan soil. U.S. Dep. Int., Off. Water Res. and Technol., Completion Rep. 14-31-0001-5085, Washington, DC. 60 p.

12. Thomas, G. W., and R. E. Phillips. 1979. Consequences of water movement in macropores. J. Environ. Qual. 8:149–152.

13. Wild, A. 1972. Nitrate leaching under bare fallow at a site in Northern Nigeria. J. Soil Sci. 23:315–324.

Chapter 8

Predicting Tillage Effects on Evaporation from the Soil[1]

D. R. LINDEN[2]

ABSTRACT

Evaporation is a major component of the water loss from the soil and has been shown to be affected by tillage and the soil conditions resulting from tillage. Enhanced water conservation is possible through the proper choice of timing and type of tillage operations. A model for predicting the evaporation of water from the soil is presented which includes these effects of tillage: the mixing and redistribution of water caused by the tillage tool, the increased radiant energy available for evaporation on roughened soil surfaces, the increased water vapor transfer away from the soil caused by increased surface area, and the water flux to the soil surface influenced by the soil hydraulic properties. Tillage-induced roughness and soil bulk density are the primary tillage inputs to the model. Model output indicates that higher evaporation losses of water stored in the tilled zone can be expected from tilled soils but that water below the tilled layer may not evaporate as freely and that tillage during an evaporation cycle will help conserve water. Evaporation predictions would be highly dependent upon infiltration predictions because the depth distribution of water has a large influence on subsequent evaporation.

[1] Contribution from the Soil and Water Management Research Unit North Central Region, USDA-ARS, St. Paul, Minn., in cooperation with the Minnesota Agric. Exp. Stn. Paper No. 11524, Scientific Journal Series.

[2] Soil scientist, USDA-ARS and assistant professor, Univ. of Minnesota, St. Paul, MN 55108.

INTRODUCTION

Evaporation is the process by which water is returned to the atmosphere from the soil at the soil surface. It is affected by the energy available to heat and vaporize water, the ease with which the vapor can move away from the soil, and the ease with which water will move to the evaporating surface from within the soil. Evaporation from a freshly wetted soil generally is limited by the available energy and aerodynamic transfer conditions (energy limiting or constant rate phase) until the surface soil becomes dry so that water flow within the soil will not meet the atmospheric demands (soil limiting or falling rate phase). Describing or predicting evaporation thus involves describing water flow within the soil, with special care being given to the boundary condition at the soil surface. Tillage affects soil water flow properties and the surface boundary conditions, and thus evaporation.

Tillage is a disturbance of soil from the surface downward to some depth. The depth of tillage depends upon the objective of the disturbance. Shallow tillage is used to control weeds, prepare seedbeds, to place seeds within the soil, to mix an additive to the soil, to change the shape of the soil-atmosphere boundary, or to change the properties of the soil affecting heat and water fluxes. Deep tillage may be used for any of the above reasons and to break up an impeding layer within the soil. This paper is concerned with describing the effects of these tillage disturbances on evaporation. The effects of closely associated phenomena, such as crop residue condition, are not considered. Evaporation rather than evapotranspiration is considered because tillage affects the surface boundary and the soil properties and thus directly affects evaporation, while in many ways tillage affects transpiration more indirectly, for example by changing the depth distribution of plant roots, crop stand, or weed populations.

Tillage has been shown to increase short-term evaporation rates (Allmaras et al., 1977; Holmes et al., 1960) and to decrease longer term evaporation (Gill et al., 1977; Willis and Bond, 1971). Although these effects are somewhat contradictory, they are consistant with some popular concepts about the use of tillage to control evaporation. In humid regions, tillage is considered a mechanism for enhancing evaporation and thus speeding soil drying. In more arid regions, tillage is considered a mechanism for water conservation. The contradictory nature of these concepts clearly establishes the need for predictive capability. Soil, climate, the timing of tillage operations, the type of tillage tool, and the depth of tillage are interactive factors that affect the soil water system. A model of evaporation considering short and long-term effects can help in making tillage decisions on a universal basis, which would be especially useful where the balance between short and long-term effects varies with the season.

Methods of predicting evaporation range from largely statistical approaches to theoretical soil water and atmospheric flow systems (Gardner, 1973; van Bavel, 1966; van Bavel and Hillel, 1976). The statistical approaches, generally, cannot be adapted for variations in soil condition and are thus not useful for the purposes of this paper. Soil water flow approaches can, however, be adapted for variations in soil conditions which are caused by tillage.

Evaporation from a soil surface is generally an unsaturated, non-steady state process that can be described by common principles of soil water movement. The literature abounds with results from the investigations of such flow systems. Variations in the form, the applicable equation, the methods of obtaining solutions, the handling of the boundary conditions, and the methods of defining the soil hydraulic properties are numerous. The author does not intend to review, discuss, and judge the merits of various existing approaches but rather to present some concepts on how any soil water model might be adapted to tillage conditions.

A review of model and experimental systems revealed several tillage-related results. First, a layer of low conductivity, such as a dry soil mulch, reduces evaporation (Hillel et al., 1975). A material with low hydraulic conductivity when unsaturated reduces the amount of water that can reach the soil surface to evaporate. The hydraulic conductivity of the surface layer is thus of critical importance to the soil limiting phase of evaporation but may also affect the energy limiting phase through effects on redistribution (Gardner et al., 1970). Secondly, structural features of a soil such as cracks (Adams et al., 1969) and variations in clod size distribution (Holmes et al., 1960) influence evaporation by affecting the movement of water vapor within the soil. Soils with high porosity or with a few large pores enhance vapor movement both by diffusion and by mass flow of air. Thirdly, the distribution of water at the start of an evaporation cycle, which can be affected by the amount and intensity of rainfall (Bresler and Kemper, 1970; Gardner and Gardner, 1969) or by properties of the soil (Ehlers, 1975; Quisenberry and Phillips, 1976) affects soil water evaporation. Water that is displaced deeper into the soil profile does not evaporate as freely as shallow water. Amount and distribution of infiltrated water is thus another critical aspect in correctly describing evaporation.

Many evaporation models are adequate predictors but lack sensitivity to tillage variables. These models could be made sensitive to tillage if hydraulic properties, the initial conditions, and the soil surface boundary conditions included the effects of tillage. Data on hydraulic properties of tilled soil are, however, rather scarce. Thus, the objective of this paper is not to present an entirely new model (which would actually not be new) but rather to present ideas, approaches, and results which can be applied to many existing models to make them sensitive to tillage conditions. The development will utilize a known soil water model (Dutt et al., 1972; Hanks and Bowers, 1962), with a few modifications as required and will consider the boundary and initial conditions which are affected by tillage. The model includes these effects of tillage: (1) changes in soil hydraulic properties resulting from changes in bulk density, (2) mixing of the soil during tillage, (3) initial conditions resulting from variations in infiltration caused by tillage, (4) soil-atmosphere boundary effects on evaporation energy resulting from radiation trapping, and (5) soil surface-atmosphere boundary contact area effects on the soil-limiting phase of evaporation.

SOIL WATER FLOW MODEL

The water flow model used was essentially the model reported by Dutt et al. (1972). It is a tridiagonal-allgorithim-solution of a finite-difference-form of the flow equation written in diffusivity and water content gradient terms for layered soils (Hanks and Bowers, 1962). The model is an isothermal model with no hysteresis in soil water properties. The hydraulic conductivity and diffusivity are expressed as a function of water content according to Campbell's (1974) procedure. The features modified to aid development of a tillage-evaporation model were the upper boundary written as the energy limiting flux and a psuedo-two-dimensional system for simulating the affects of increased soil-atmospheric contact area caused by roughness (cloddiness) of the surface. The soil profile was considered deep enough so that the lower boundary condition could be expressed as a constant water content with no water flux. The model is operated with a daily equivalent potential evaporation rate as the surface layer flux during early evaporation. The model outputs are evaporation rates and amounts, drainage rates and amounts, and water contents and storage at specified time intervals.

HYDRAULIC PROPERTIES

Data on the relationship between conductivity (K), diffusivity (D), water content (θ), and matric potential (ψ) in freshly tilled soils are practically nonexistant. Experimentally, these relationships are very difficult to measure since the first introduction of water begins to change the properties. In the absence of such data, a first approximation technique utilizing similarities of desorption curves (ψ vs. θ) for a given soil at different porosities supplied input parameters to the model.

Primary tillage generally increases total porosity (α) by increasing the large pores within the soil (Van Duin, 1956). Water content (θ) and matric potential (ψ) relationships are thus most affected by tillage at ψ greater than about -0.1 bar (wet end of the water content scale). At ψ less than about -0.1 bar, ψ vs. water contents on a weight basis (W) relationship, (ψ vs. W curves) for a given soil are nearly independent of porosity (Van Duin, 1956). Thus, at ψ less than about -0.1 bar a relationship between ψ and θ for a soil at a given porosity could be used to approximate a ψ vs. θ relationship at another porosity by assuming that ψ vs. W is a unique function (does not vary with porosity) for soil. This is, an approximation of course, since ψ vs. W curves at different porosities do not exactly coalesce (Van Duin, 1956). This approximation is not valid when matric potentials are greater than about -0.1 bar which generally occurs only briefly during and immediately after soil wetting. This period, for purposes of this model, can be handled separately as part of infiltration and initial water contents distribution. The upper and lower θ limits for this approximation to be valid were calculated from W valves at -0.1 and -15.0 bar matric potentials and the bulk density of the soil. Matric potential vs. volumetric water content curves were inputs to the model for calculating D vs. θ relationships.

Campbell's (1974) procedure, which is consistent with Brooks and Corey's (1964) development, was selected as a possible method of describing K vs. θ and D vs. θ relationships. Hydraulic properties as functions of volumetric water content were input to the model as a function of bulk density by using a desorption slope parameter (B), maximum hydraulic conductivity, and air entry matric potential taken from literature (Moore et al., 1980) all of which did not vary with bulk density and by computing the upper and lower volumetric water content limits from water contents by weight which also was not affected by bulk density. The preceding discussions are illustrated in Fig. 1, which shows the approximate conduc-

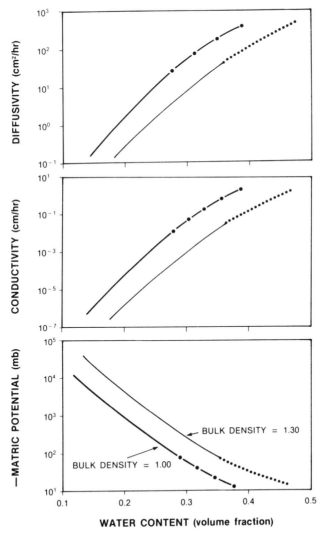

Fig. 1. Matric potentials, hydraulic conductivity, and soil water diffusivity as functions of volumetric water contents for two bulk densities used as model input.

tivity and diffusivity vs. θ relationships used as input for the solution of the evaporation problems. Improvements in these relationships are quite possible through experimental data.

Simulation runs to illustrate the possible effects of decreased bulk density caused by tillage on evaporation were made for two hypothetical soil profiles. The two profiles were a uniform soil profile with a bulk density of 1.3 g/cm³ and a layered soil. The layered soil was simulated with a 26-cm deep layer with a bulk density 1.0 g/cm³ overlaying the 1.3 g/cm³ bulk density soil below. A 26-cm layer of 1.0 g/cm³ bulk density soil could have been created by tilling the 1.3 g/cm³ soil to a 20 cm depth.

The initial volumetric water contents near the soil surface for a simulated evaporation period for the uniform soil profile were the maximum water contents of the soil during infiltration, which has been reported as approximately 85% of the total porosity (Jackson, 1963). It was assumed that the initial volumetric water contents in the lower density soil of the layered profile would be equal to the water contents by weight of the high bulk density soil times the bulk density since higher water contents exist only temporarily and can be accounted for by nearly instantaneous redistribution of water. This assumption is invalid, of course, if an impeding layer restricts the rapid displacement of this water. The initial

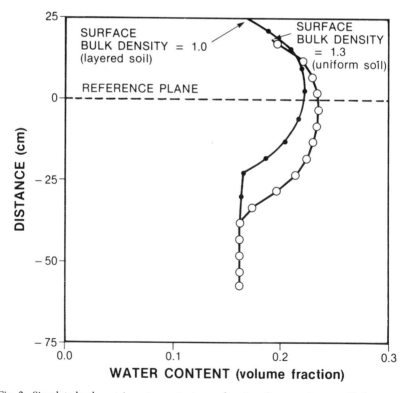

Fig. 2. Simulated volumetric water contents as a function distance above and below a reference plane for two soils at different bulk densities above the reference plane and the same bulk density below the plane. Hydraulic property inputs are shown in Fig. 1.

water contents for both soils were the maximum water content to a depth sufficient to have a total water storage 5 cm more than the water storage at 15-bar water content. The initial water contents below this depth were the 15-bar water contents. All other inputs including the potential evaporation rate were the same for both simulations.

   Predicted water content (θ) vs. depth profiles after 10 days of redistribution and evaporations are shown in Fig. 2 for the two hypothetical soils. Predicted cumulative evaporation vs. time for the same two hypothetical soil profiles are shown in Fig. 3. Evaporation rates (Fig. 3) were higher in the soil with a low bulk density surface layer than the soil with a high bulk density surface layer time after the initial constant rate period but eventually they became nearly equal. There is a difference of only 0.2 cm water storage in the profiles after 10 days of evaporation (Fig. 2), however, the distribution of water relative to the reference plane (a reference plane at 20 cm depth in the uniform profile was selected because there is an equal mass of soil above this plane in both soil profiles) is quite different. There is considerably more water below the reference plane in the high bulk density surface soil, which indicates that over extended periods of time the high bulk density surface would probably lose con-

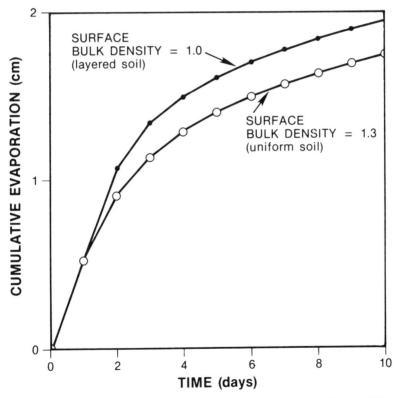

Fig. 3. Simulated cumulative evaporation as a function of time for soils with two different bulk densities in the surface layer (See Fig. 2). Hydraulic property inputs are shown in Fig. 1.

siderably less water by evaporation than the low bulk density surface. These limited simulations should not be related to actual field soil water regimes however, since the effects of infiltration and other evaporation modifying factors have not yet been included.

INFILTRATION AND INITIAL WATER CONTENT DISTRIBUTION

Tillage has a large influence on infiltration and the distribution of infiltrated water within the soil, as was previously discussed in this symposium. Macroporous flow, from both worm holes and cracks in an untilled, noncrusted soil and the large pores of freshly tilled soils, contribute to a deeper displacement of water. The deeper that water is displaced into the soil, by whatever means, the less free it is to evaporate (Bresler et al., 1969). Existing models can predict evaporation, when the initial water contents are known, and therefore, will not be discussed here. This omission does not imply, however, that infiltration can be separated from evaporation, one has a large influence upon the other.

TILLAGE DURING EVAPORATION

Practical applications of an evaporation model must provide for tillage events to interrupt an otherwise continuous process. In modeling, terminology tillage can be considered a disturbance that instantly changes the conditions. Tillage causes an immediate change in the hydraulic properties, the soil-atmosphere boundary, and instantaneously redistributes water by mechanical means. Tillage can be assumed to cause a complete mix of the soil which means that water can be instantly redistributed uniformly over the tilled depth.

The results of simulated tillage on evaporation from the soil surface are shown in Fig. 4. The initial conditions for these simulations were identical to the previous (Fig. 3) uniform profile case. Tillage was simulated after 1 or 5 days of evaporation by (1) instantly redistributing the total water contained in the top 20 cm of the profile uniformly over the resultant (after tillage) 26 cm depth and (2) decreasing the bulk density from 1.3 g/cm^3 to 1.0 g/cm^3 in this layer, thus, creating a layered profile case. The evaporation process was then restarted with the modified properties and initial water contents of the surface layer. Because evaporation and redistribution had reduced the water storage in the top 20 cm of soil profile (20 cm depth prior to tillage), the initial water contents for the restart of evaporation were less than the initial water contents of the layered profile case shown earlier (Fig. 3). The soil surface, thus, dried quickly, and resulted in lower evaporation rates from the tillage-layered soil. The effects of time of tillage, observed in experimental data (Gill et al., 1977; Willis and Bond, 1971), were reproduced by this model simply by interrupting the continuous process and instantaneously changing the properties and the water contents.

POTENTIAL EVAPORATION FLUX

The potential rate is strongly influenced by albedo of the soil (Idso et al., 1975; van Bavel and Hillel, 1976). Roughness reduces the albedo of a soil (Allmaras et al., 1972; Bowers and Hanks, 1965; Cary and Evans, 1975; Gausman et al., 1977). In 1978, the author presented the concept and a model to account for these reduced albedos by the mechanism of radiation trapping on roughened soil surfaces (Cruse et al., 1980; Linden, 1978; Linden, 1979[3]). Briefly, the model assumes that reflected sunlight or emitted terrestrial radiation is perfectly scattered and that variations in topography (roughness) can, therefore, intercept this radiation so that it does not escape from the soil system into the atmosphere. Microtopographic data and an albedo for a smooth condition can be used to estimate

---

[3] Linden, D. R. 1979. A model to predict soil water storage as affected by tillage practices. Ph.D. Thesis. Dep. of Soil Science, Univ. of Minnesota, St. Paul, Minn.

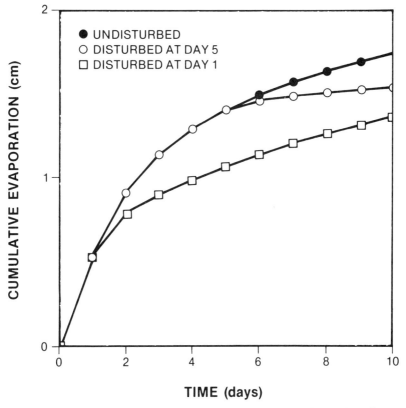

Fig. 4. Simulated cumulative evaporation as a function of time for soils disturbed after 1 and 5 days of evaporation and an undisturbed soil. Disturbance was simulated by decreasing the bulk density from 1.3 to 1.0 at the designated time which thereby changing the hydraulic properties (Fig. 1) used as model input and by instantly redistributing the water in the disturbed zone.

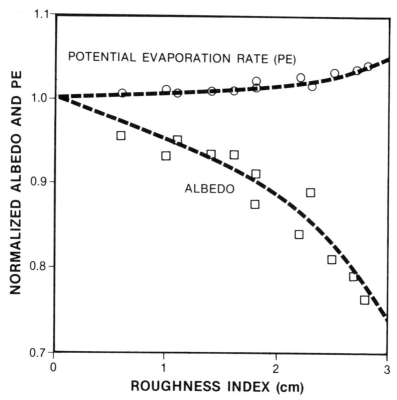

Fig. 5. Predicted relative albedo and potential evaporation rate (PE) as a function of roughness index (standard deviation for elevation measurements). Data shown is relative to
values for a smooth surface (roughness index = 0).

the magnitude of radiation trapping and albedo on the rough soil. The
model was tested by comparing measured and predicted albedo for a 4-
hour midday period on a rough soil when the soil surface was dry. Albedo
of a smooth tilled soil was measured to be 0.179 for use in predicting the
albedo of the rough surface. Albedo for the rough surface was obtained
from an upright and an inverted Epply[4] pyranometers. The predicted
albedo was 0.131 and the measured was 0.130.

Since the model appeared to give reasonable predictions of albedo a
general relationship between roughness and albedo was developed.

Several sets of microtopography data with various roughness conditions were subjected to model calculations and the predicted albedos were
correlated with the standard deviation (SD) of the microtopographic
data. The standard deviation of elevations (σ) is used as an index of the
roughness with higher σ occurring on rougher surfaces. Predicted albedos
relative to a smooth condition plotted as a function of roughness index are
shown in Fig. 5. As roughness increases, the albedo decreases (Fig. 5)

[4] Company or trade names are mentioned for the reader's benefit and do not imply prefer-
ence by the USDA over products not mentioned.

which will tend to increase the potential evaporation rate (PE). For purposes of illustrating this effect of tillage on evaporation, predicted PE's relative to a smooth condition with roughness affected albedo as the only variable input are also shown in Fig. 5. The PE's (Fig. 5) were estimated with an equation based on radiant energy (Van Doren and Allmaras, 1978) using reasonable and nonvariant assumptions for the remaining equation terms. Roughness related albedo did not have a large impact on the potential evaporation rate (Fig. 5), however, small differences in this rate can have a large impact on evaporation over long time periods.

TRANSITION TO SOIL-LIMITING EVAPORATION RATE

Tillage affects roughness and therefore the area of the soil directly in contact with the atmosphere. Some reasonable speculation would indicate that when the limiting evaporation flux is just below the soil surface, increased soil-atmosphere contact area would increase the evaporation rate because of a horizontal water flux component. As evaporation proceeds, the limiting flux becomes vertical and would thus not be

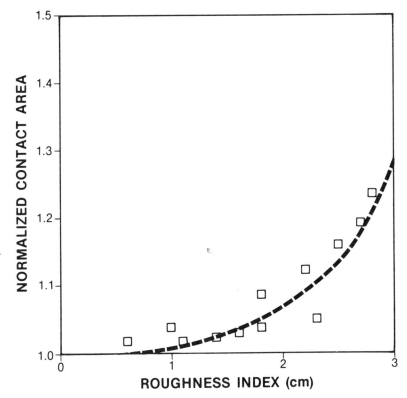

Fig. 6. Soil-Atmosphere contact area per unit horizontal area as a function of roughness index.

affected by soil-atmosphere contact area. The initial phase of evaporation is limited by atmospheric conditions and is thus also unaffected by soil-atmosphere contact area. The period between atmospheric limited and soil-vertical-flux limited evaporation rate is a transition period and would be affected by soil-atmosphere contact area. Relative contact areas (calculated from microtopographic data) are plotted as a function roughness index in Fig. 6. Contact areas were calculated as the sum of lengths of straight line segments per unit horizontal distance between microrelief measurement points. Surface areas 1.3 times the horizontal area are possible on roughly tilled lands.

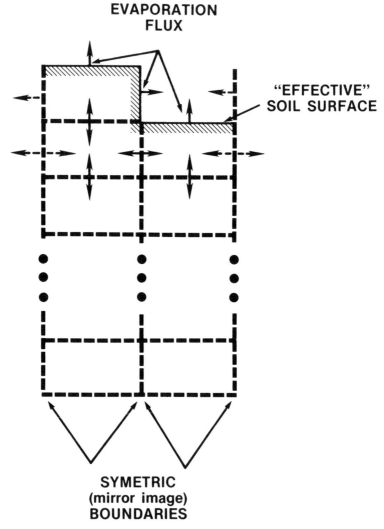

Fig. 7. Schematic representation of two-dimensional flow system, showing the elevated soil surface for one of two segments and the mirror image symmetry at the horizontal boundaries. See text for discussion of this model.

The effect of soil-atmosphere contact area on evaporation was modeled by modifying the water flow model for horizontal flow. Two adjacent vertical flow models (Fig. 7) are considered simultaneously. When evaporation is limited by the soil, a horizontal component of flux is added to the evaporation rate. Horizontal evaporation flux is a function of soil-atmosphere contact area. Horizontal flow at all depths within the soil is also calculated between time steps for vertical flow. A two-segment horizontal flow model is sufficient because tillage roughness is essentially random in nature (not considering ridges and furrows so that mirror-image-symmetry can be assumed at the horizontal boundaries.

Simulated evaporation as a function of time for two roughness conditions are shown in Fig. 8. The initial and soil profile conditions for the simulations were identical to the layered soil profile shown previously (Fig. 3). The only difference between these simulations was roughness and associated contact area. All other conditions were equal for both simulations. The roughness index was 3.0 cm and the relative contact area was 1.25 for the simulated rough soil. While the smooth soil had a roughness index of zero and a relative contact area of 1.0. Predicted evaporation rates were initially higher on the rough surface because of the reduced albedo. Evaporation rates remained higher on the rough surface, as

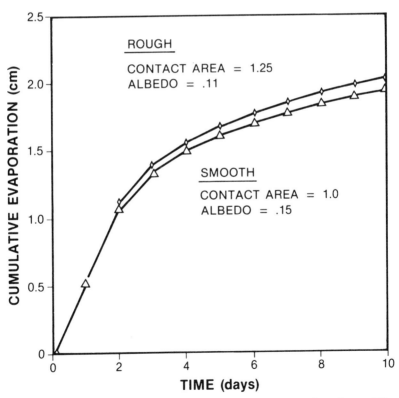

Fig. 8. Simulated cumulative evaporation as a function of time for soils with two different roughness indexes and soil-atmosphere contact areas.

the rates on both soils began to fall, because of the increased soil-atmosphere contact area and the horizontal component of flow. Evaporation rates, however, became nearly equal (Fig. 8) at longer times because the limiting flux was primarily vertical. The magnitude of the effect of contact area on evaporation rate appears to be smaller than anticipated from the preceding discussion because the horizontal evaporation component reduces the subsequent vertical component, thus, causing only a small net increase in evaporation.

TILLAGE EFFECT ON EVAPORATION

The combined effects of bulk density affected hydraulic properties and roughness affected radiation trapping and contact area on evaporation can be surmised from the previous discussion. Tillage to a 20 cm depth, as simulated by these three characteristics, under ideal initial conditions would increase evaporation of water stored on the tilled zone, but because of rapid surface drying, could result in conservation of water stored below this depth. These simulations however, are difficult to relate to water storage under tillage systems in the field, since the initial conditions were rather idealistic and the complicating feature of differences in infiltration were ignored. They do, however, illustrate that tillage will affect evaporation by its effect on liquid water flow properties and the surface boundary conditions, and that simulations of the process with simple data inputs are possible.

SUMMARY AND CONCLUSIONS

In summary, the model presented here would require as the minimum input information the following: (1) a set of mass water content vs. matric potential points covering the range of water contents encountered in the field, (2) the bulk density of the soil profile, (3) a maximum hydraulic conductivity, (4) a roughness index, and (5) albedo of the soil in a smooth dry condition. Simulation results would be improved if, in addition to the minimum data input, these data were also provided: (1) volumetric water content vs. matric potential at the density of soil to be simulated, (2) hydraulic conductivity or diffusivity as a function of water content or matric potential, (3) hydraulic properties of the entire soil profile, (4) albedo of the soil, and (5) surface area of soil in direct contact with the atmosphere. The output of the model includes the evaporation rate, evaporation amount, soil water storage, and water contents at specific times.

The major benefit from modeling evaporation has been the isolation and study of single variables affected by tillage. These isolation studies have shown that roughness is of some significance, but that the hydraulic properties of the soil are of major concern. Other affects of tillage still need to be considered in order for a model to be completely sensitive to tillage. The effect of roughness on the aerodynamic transfer of water vapor away from the soil surface, the effect of roughness on nonuniform drying (van Bavel and Hillel, 1976) caused by nonuniform radiation inputs, the effect of roughness and porosity on thermal gradients which induce water flow, and the effects of porosity on water vapor movement through diffusion and mass air movement should be included.

LITERATURE CITED

1. Adams, J. E., J. T. Ritchie, E. Burnett, and D. W. Fryrear. 1969. Evaporation from a simulated soil shrinkage crack. Soil Sci. Soc. Am. J. 33(4):609–613.

2. Allmaras, R. R., E. A. Hallauer, W. W. Nelson, and S. D. Evans. 1977. Surface energy balance and soil thermal property modifications by tillage-induced soil structure. Minnesota Agric. Exp. Stn. Tech. Bull. 306. 40 p.

3. ————, W. W. Nelson, and E. A. Hallauer. 1972. Fall vs. spring plowing and related soil heat balance in the western corn belt. Minnesota Agric. Exp. Stn. Tech. Bull. 283. 22 p.

4. Bowers, S. A., and R. J. Hanks. 1965. Reflection of radiant energy from soils. Soil Sci. 100(2):130–138.

5. Bresler, E., and W. E. Kemper. 1970. Soil water evaporation as affected by wetting method and crust formation. Soil Sci. Soc. Am. J. 34(1):3–8.

6. ————, ————, and R. J. Hanks. 1969. Infiltration, redistribution, and subsequent evaporation of water from soil as affected by wetting rate and hysteresis. Soil Sci. Soc. Am. J. 36(6):832–840.

7. Brooks, R. H., and A. T. Corey. 1964. Hydraulic properties of porous media. Colorado State University, Fort Collins, Colorado, Hydrology paper 3:1–27.

8. Campbell, G. S. 1974. A simple method for determining unsaturated conductivity from moisture retention data. Soil Sci. 117(6):311–314.

9. Cary, J. W., and D. D. Evans (ed.). 1975. Soil crusts. Arizona Agric. Exp. Stn. Tech. Bull. 214.

10. Cruse, R. M., D. R. Linden, J. K. Radke, W. E. Larson, and K. Larntz. 1980. A model to predict tillage effects on soil temperature. Soil Sci. Soc. Am. J. 44(2):378–383.

11. Dutt, G. R., M. J. Shaffer, and W. J. Moore. 1972. Computer simulation model of dynamic bio-physicochemical processes in soil. Arizona Agric. Exp. Stn. Tech. Bull. 196. 99 p.

12. Ehlers, W. 1975. Observation on earthworm channels and infiltration of tilled and untilled loose soil. Soil Sci. 119(3):242–249.

13. Gardner, H. R. 1973. Prediction of evaporation from homogenous soil based on the flow equation. Soil Sci. Soc. Am. J. 37(4):513–516.

14. ————, and W. R. Gardner. 1969. Relation of water application to evaporation and storage of soil water. Soil Sci. Soc. Am. J. 33(2):192–196.

15. Gardner, W. R., D. Hillel, and Y. Benyamini. 1970. Post-irrigation movement of soil water. 2. Simultaneous redistribution and evaporation. Water Resour. Res. 6(4):1148–1153.

16. Gausman, H. W., R. W. Leamer, J. R. Noreiga, R. R. Rodrigeuz, and C. L. Wiegand. 1977. Field measured spectroradiometric reflections of disked and nondisked soil with and without wheat straw. Soil Sci. Soc. Am. J. 41:793–796.

17. Gill, K. S., S. K. Jalota, S. S. Prihar, and T. N. Chauphary. 1977. Water conservation by soil mulch in relation to soil type, time of tillage, tilth and evaporativity. J. Indian Soc. Soil Sci. 25(4):360–366.

18. Hanks, R. J., and S. A. Bowers. 1962. Numerical solution of the moisture flow equation for infiltration into layered soils. Soil Sci. Soc. Am. Proc. 26:530–534.

19. Hillel, D. I., C. H. M. van Bavel, and H. Talpaz. 1975. Dynamic simulation of water storage in fallow soil as affected by mulch of hydrophobic aggregates. Soil Sci. Soc. Am. Proc. 39:826–833.

20. Holmes, J. W., E. L. Greacen, and C. G. Gurr. 1960. The evaporation of water from bare soils with different tilths. Int. Congr. Soil Sci. Trans. 7th (Madison, Wis.) II:188–194.

21. Idso, S. B., R. D. Jackson, and R. J. Reginato. 1975. Estimating evaporation: A technique adaptable to remote sensing. Science 189:991–992.

22. Jackson, R. D. 1963. Porosity and soil water diffusivity relation. Soil Sci. Soc. Am. Proc. 27(2):123–126.

23. Linden, D. R. 1978. A model of multiple reflection absorbance of solar radiation on tillage-induced rough soil surfaces. Agron. Abstr. Am. Soc. of Agron., Madison, Wis. p. 202.

24. Moore, I. A., C. L. Larson, and D. C. Slack. 1980. Predicting infiltration and micro-relief surface storage for cultivated soils. Univ. of Minn. Graduate School WRRC Bull. 102. 122 p.

25. Quisenberry, V. L., and R. E. Phillips. 1976. Precolation of surface applied water in the field. Soil Sci. Soc. Am. J. 40(4):484–489.

26. van Bavel, C. H. M. 1966. Potential evaporation: The combination concept and its experimental verification. Water Resour. Res. 2:455–467.

27. ————, and D. I. Hillel. 1976. Calculating potential and actual evaporation from a bare soil surface by simulation of concurrent flow of water and heat. Agric. Meteorol. 17:453–476.

28. Van Doren, D. M., and R. R. Allmaras. 1978. Effect of residue management practices on the soil physical environment, microclimate, and plant growth. p. 49–84. In Am. Soc. Agron. Spec. Pub. no. 31. Crop residue management systems. ASA, CSSA, SSSA, Madison, Wis.

29. Van Duin, R. H. A. 1956. On the influence of tillage on conduction of heat diffusion of air and infiltration of water in soil. Reports of Agricultural Investigations No. 62.7. The Hague, Netherlands. p. 1–109.

30. Willis, W. O., and J. J. Bond. 1971. Soil water evaporation: Reduction by simulated tillage. Soil Sci. Soc. Am. Proc. 35(4):526–529.

Chapter 9

Modeling Tillage Effects on Soil Temperature[1]

R. M. CRUSE, K. N. POTTER, AND
R. R. ALLMARAS[2]

ABSTRACT

The interaction of microclimate with the soil surface to produce a given soil temperature regime is complex and dynamic. Soil surface configuration and plant residue cover, both affected by tillage, have significant impact on soil heat flux and, therefore, soil temperature.

Strengths and weaknesses of past tillage, soil-temperature research are discussed. Advantages of modeling tillage effects on soil temperature also are presented. Soil properties known to affect soil heat flux are identified. Models from the literature for predicting soil temperature are discussed and evaluated on the basis of their sensitivity to soil property changes induced by tillage and their sensitivity to soil properties known to affect soil heat flux. It is concluded that additional modeling efforts are needed that address: 1) the heat exchange and energy balance processes at the soil surface and 2) the changes in soil surface condition that occur as a function of time, climate, and tillage events.

[1] Joint contribution: Journal Paper J-10284 of the Iowa Agric. and Home Econ. Exp. Stn., Ames, IA (Project No. 2423), and USDA, ARS.

[2] Assistant professor and graduate assistant, Agronomy Dep., Iowa State University; and soil scientist, Columbia Plateau Conserv. Res. Ctr., Pendleton, OR, respectively.

INTRODUCTION

Tillage research, which began centuries ago, has yielded mostly qualitative observations because there are poor theoretical foundations upon which to guide the collection and use of quantitative observations. This lack of theory and associated mathematical quantification between a dependent variable (i.e., soil temperature) and a tillage-created soil physical condition has limited the inferences of tillage effects on soil temperature to only the experimental conditions of the test where the observations were made. This situation has greatly limited our general knowledge of soil temperature response to tillage and the modifying effects of soil type or climate, for example. Moreover, advances in microclimatology or micrometeorology are only marginally helpful for understanding tillage effects on soil temperature because the important tillage-related soil properties have not been recognized and characterized.

While qualitative judgments about tillage have given remarkable success with tillage systems thus far, future studies must be more quantitative to assure continued success with reduced-tillage systems. It is the objective of this paper to: 1) identify research needs in the area of tillage and soil temperature, 2) discuss the advantages of modeling tillage effects on soil temperature, 3) examine existing soil heat flux models to determine which soil properties and soil-influenced aerodynamic factors significantly affect soil heat flux, and 4) examine the sensitivity of existing soil temperature models to those soil properties and aerodynamic factors altered by tillage.

Tillage-Soil Temperature Interaction

Plant growth and crop yields integrate the effects of soil aeration, water content, temperature, nutrition, and phytotoxic compounds. To draw valid conclusions about the effects of one factor (soil temperature) on yield, all other environmental factors across tillage treatments must be: 1) kept constant, 2) monitored, or 3) assumed nonlimiting. Soil conditions created by different tillage practices affect each of many soil environmental factors; thus, each soil environmental condition cannot be considered constant across tillage treatments. No work has been located that has monitored all conditions listed. Thus, to draw conclusions about tillage-induced soil temperature effects on yield, one must assume that all other soil environmental conditions are nonlimiting, a condition that seldom exists throughout the growing season under field conditions. It thus appears that research should concentrate on soil temperature response to soil physical property effects, rather than on tillage implement or practice effects.

To understand how tillage affects soil temperature, one must understand the heat exchange process between the atmosphere and soil and how this exchange process is influenced by both soil and micrometeorological conditions. To illustrate, micrometeorology may be viewed as the driving mechanism, while the soil surface layer (including a tilled layer) may be viewed as a filter, modifying the effect of micrometeorological inputs on soil temperature. The problem often is more complex because the soil surface physical condition, the filter, is dynamic. Changes occur as a function of time and space because of meteorological events, tillage events, and biological activity. Not until work of Larson (1964) was there emphasis placed on the soil condition or property created by the tillage implement(s) and on the micrometeorological factors interacting with the soil condition to produce a characteristic soil temperature regime. This approach must guide future research efforts if there are to be meaningful responses to national goals (Bertrand, 1980) of energy and soil conservation with reduced tillage.

These circumstances (i.e., the dynamic nature of the surface and the interacting processes) suggest that modeling is an advantageous approach for studying tillage effects on soil temperature. Model development promotes identification of individual processes or components within the complex soil heat-exchange, soil-temperature system. These components can then be studied individually or in conjunction with other components or processes. Sensitivity analysis in a model is sometimes the only way to study the effect of tillage-induced soil conditions on each of the processes relating to soil heat exchange and soil temperature. Model development also encourages organization and incorporation of existing data into a unifying conceptual framework. These approaches should maximize use of existing data and should increase effectiveness of future experimental plans made in accord with theory.

Generalized Concepts of Energy Balance and Soil Heating

The temperature of any position within the soil is controlled by the quantity of energy in the system acting to influence temperature, the distribution of heat energy within the body, and heat capacity of the media. This can be summarized mathematically as:

$$Q = -A \lambda (\partial T)/\partial Z \qquad [1]$$

and

$$C = \partial Q/\partial T \qquad [2]$$

where
   Q = heat flux (W)
   A = cross-sectional area (cm^2)
   λ = thermal conductivity (W cm^{-1} C^{-1})
   T = temperature (C)
   Z = distance (cm)
   C = heat capacity (J C^{-1})
Tillage alters λ and C because of effects on the spatial distribution of soil components, the volumetric proportion of each soil component at each position, and orientation of each soil component (de Vries, 1963). These soil properties exert an influence on heat entering the soil and on subsequent movement in the soil. Simulations, such as those by Hanks et al.

(1971) and Wierenga and de Wit (1970), and detailed measurements, such as those by Allmaras et al. (1977), indicate that these thermal properties may be estimated with sufficient accuracy for temperature modeling purposes.

Future advances in the modeling of soil temperature will depend upon our ability to simultaneously improve measurement and modeling of soil heat flux at the soil surface. This flux is, in turn, related to the energy balance at the surface. Soil heat flux density can be presented as the residual in an energy flux density balance (in which H_g can be determined if R_n, H_a, and L_e are known):

$$H_g = R_n - (H_a + L_e) \qquad [3]$$

where

H_g = net energy flux density available for soil warming (W m^{-2})
R_n = net radiation flux density, includes longwave and shortwave radiation (W m^{-2})
L_e = latent energy flux density (W m^{-2})
H_a = sensible heat energy flux density conducted to the atmosphere (W m^{-2})

In modeling H_g, two approaches can be taken. Either the right side of Eq. [3] can be solved for each of three components or H_g can be calculated directly from soil measurements with only minor consideration for atmospheric conditions.

Models or methods of calculating each component on the right-hand side of Eq. [3] are required if the energy balance approach is used to obtain H_g indirectly. The R_n can be estimated by solving the following equation:

$$R_n = (1 - \alpha_s)\, R_s + R_l - \varsigma\sigma\, (T_0 + 273.16)^4 \qquad [4]$$

where

α_s = soil reflection coefficient for shortwave radiation
R_s = flux density of incoming shortwave radiation (W m^{-2})
R_l = flux density of incoming longwave radiation (W m^{-2})
ς = surface emittance
σ = Stefans constant (W m^{-2} K^{-4})
T_0 = surface soil temperature (C)

The R_s can be computed (List, 1968) from meteorological data, but some of the required parameters are not readily estimated, and, if short-term variations are necessary, they will not be depicted. Fortunately, R_s often is measured at weather shelter height. Estimation of α_s can be by methods of Linden (1979)[3], Cruse et al. (1980), or van Bavel and Hillel (1976).

All three methods estimate α_s for a smooth soil condition at the specified soil water content. Linden (1979)[3] and Cruse et al. (1980) then adjust the value of α_s for surface-roughness effects on radiation absorption. Tillage effects on soil surface roughness therefore are projected to influence

---

[3] Linden, D. R. 1979. A model to predict soil water storage as affected by tillage practices. Ph.D. Thesis. Univ. of Minnesota. St. Paul, Minn.

α_s. Linden (1981) indicated that, in going from a smooth surface to a very rough soil surface condition, as might be induced by primary tillage, α_s could vary by as much as 25%. The R_l could be calculated from Brunt's formula (Penman, 1948) as a function of air temperature and dew point temperature. Idso and Jackson (1969) and Brutsaert (1975) also have computed R_l. Linden (1981) has simulated L_e for the tilled soil condition.

Since our ultimate goal is to predict soil temperature, we will not have the convenience of knowing the soil surface temperature, T_0 in Eq. [4]. Thus, a technique such as that developed by Gupta et al. (1981), which is based on air temperature and percentage residue cover, would be required for soil surface temperature estimation. Approximately 95% of their hourly predictions were within $\pm$ 10.0 and 4.5°C for bare and residue-covered soil surfaces, respectively. Assuming an emissivity value of 0.9 for the soil surface and a soil temperature of around 25 C, a 5 to 10 C error in surface temperature estimation would yield calculated longwave emission errors of 26 to 51 W m^{-2}, respectively. A problem arises in the use of this technique (Gupta et al., 1981) for surface temperature and, thus, longwave emission calculations for residue-covered soils. As surface cover increases, the residue becomes the dominant emitter of outgoing radiation. This technique of estimating soil surface temperature does not estimate residue temperature; thus, the required information for calculating residue emission is not available.

The remaining factor on the right-hand side of Eq. [3] that requires estimation is H_a. This factor is very sensitive to the air, soil-temperature differential, the surface shape, and turbulence of air moving across the surface (Sears and Zemansky, 1970). A means of suitably predicting H_a for tilled soil surfaces under field conditions has not yet been identified.

Field Measurement of Soil Heat Flux in Tillage Studies

The difficulty of measuring or modeling R_n, H_a, and L_e in Eq. [3] has prompted investigators to measure or model soil temperature (and/or H_g) directly unless there was some other need to have independent estimates of R_n, H_a, or L_e. Examples of this direct approach are evident in the measurements of Allmaras et al. (1972, 1977) and the models of Gupta et al. (1981).

Another common field approach to soil heat flux determination is to work with the energy balance in the form:

$$H_a + L_e = R_n - H_g. \qquad [5]$$

In this form, emphasis is placed on the ease of measuring R_n and H_g and the difficulty of separating H_a and L_e. The most accurate field means for determining H_a and L_e separately is the eddy correlation method while the energy balance method is most popular by utilizing the Bowen ratio (Tanner, 1968). While it is recognized that H_g in Eq. [3] is readily measured, it often is approximated (Jensen, 1974) as:

$$H_g = 100 \, (T_a^{n+1} - T_a^{n-1})/\Delta t \qquad [6]$$

where T_a is daily average air temperature for Day n, and t is time in days. Justification for this approximation is that net H_g on a daily basis is small, H_g has a small amplitude under a crop canopy or mulch, and soil temperature over a long period changes as the air temperature changes. Another common approximation is to assume that mean H_g is zero if there is no net heating or cooling in a 24-hour period such as is done in remote sensing of soil moisture and evaporation (Price, 1980) even though surface temperatures affect the thermal infrared radiation emitted. Consequently, the literature contains little or no soil heat flux modeling efforts and only limited soil heat flux measuring efforts. Moreover, when soil heat flux is large and subject to tillage management in the absence of full plant canopy, the direct measurement of soil heat flux is difficult because of the ill defined soil-air interface. This is particularly true when a soil is rough and/or covered with a plant residue.

In any effort to model soil heat flux, the models must be tested under field conditions. It is therefore helpful to review progress and the present state of the science for measuring soil heat flux. There are four methods for directly measuring soil heat flux. These methods are not easily modeled, but they are needed to verify soil temperature models.

Vertical heat flux, downward through a very thin soil surface layer, may be expressed as a form of Eq. [1]:

$$H(0,t) = -\lambda[dT\,(Z,\,t)]/dZ \qquad [7]$$

H is a generalized soil heat flux density (heat passing through a unit area per unit time) for a given depth (Z) and time (t); λ is the thermal conductivity; dT/dZ is the temperature gradient. $H(0,t)$ corresponds to H_g in Eq. [3].

As will be discussed later, the thermal-gradient method (Eq. [7]) may be applied for situations where $Z = 0$. However, at the soil surface ($Z = 0$), Eq. [7] is not a useful technique because λ changes rapidly as soil moisture changes; furthermore, the temperature gradient over depth changes rapidly as a function of depth and time. This method does not account for latent heat flux.

A second method is merely to use a heat flux plate, a thermopile associated with a thin plate in the horizontal plane, as discussed by Fuchs and Tanner (1968). Heat flux plates cause heat flow to converge or diverge if their thermal conductivity differs from the soil (Phillip, 1961). They may also give unreal heat flow estimates if condensate collects at the plate-soil interface as a result of interference with liquid and vapor flows in the soil. Because of these problems heat flux plates should not be placed at depths of less than 10 cm and therefore are not satisfactory for measuring $H(0,t)$ unless used in a combination method to be discussed later. Examples of heat flux plate use at the 1-cm depth are Idso et al. (1975) and Aase and Siddoway (1980); $H(0,t)$ was not sought in these experiments. A special problem with heat flux plates within the tilled layer of tillage experiments is the inaccurate readings caused by nonhorizontal placement and soil disturbance during placement. In tillage experiments, heat flux plates should be placed no shallower than the deepest tilled layer in the experiment.

A third method of estimating soil heat flux is the calorimetric method, in which the change in soil heat storage in a very thin soil layer is equated to the temperature change over time:

$$- \partial H (Z, t)/\partial Z = C_1[\partial T (Z, t)]/\partial t \qquad [8]$$

In Eq. [8], the volumetric heat capacity is represented by C_1. This method can be used to measure $H(0,t)$. It is more useful than Eq. [7] because $\partial T/\partial t$ is much easier to measure than is $\partial T/\partial Z$ near the soil surface. Moreover, C_1 is much less sensitive to soil moisture variations than is λ (Fuchs and Hadas, 1972). Equation [8] may give poor measurements of $H(0,t)$ if there are small errors in soil temperature measurement below about the 30 cm soil depth (Kimball and Jackson, 1975). $H(0,t)$ can be measured by using the calorimetric approach exclusively with shallow (i.e., 1-cm) soil temperature measurements (Hanks and Jacobs, 1971; Hanks and Tanner, 1972). Wierenga and de Wit (1970) show estimates of $H(0,t)$ using the calorimetric approach; their shallowest temperature measurement was at the soil surface (Wierenga et al., 1969).

Numerical approximations of Eq. [8] are used commonly to estimate soil temperature at any test depth, given an upper and lower boundary soil temperature as a function of time. Upper boundary temperatures chosen are those at 1 cm (Hanks et al., 1971), 5 cm (Wierenga et al., 1969), or surface temperature (Wierenga and de Wit, 1970). This third method vividly displays the fact that, if we knew $H(0,t)$ or $T(0,t)$, the problem of modeling soil temperature would be much easier.

A fourth and most useful method of measuring H_g directly is the combination method as shown in Eq. [9a] and [9b]:

$$- \partial H(Z, t)/\partial Z = C[\partial T(Z, t)]/\partial t; \quad Z \le Z_1 \qquad [9a]$$

$$H(Z_1, t) = -\lambda[dT(Z_1, t)]/dZ \quad \text{or} \quad K_x^1 E(Z_1, t); \quad Z = Z_1 \qquad [9b]$$

where K_x^1 is the constant relating thermopile signal E to $H(Z_1, t)$. This method utilizes calorimetry at depths of $\le Z_1$, as indicated in Eq. [9a], and either heat flux plates or Eq. [7] to determine heat flux across the plane corresponding to Z_1 as indicated in Eq. [9b].

For tillage studies, this combination approach is especially useful. Only soil temperature is measured at depths $\le Z_1$, where soil disturbance is especially critical in the tilled layer. (Moisture, bulk density, and other such measurements are necessary but can be adjacent to a thermocouple network.) At Z_1, one can utilize heat flux plates because, at this depth, they are easier to keep horizontal, will cause less soil disturbance, and will interfere less with vapor and liquid flows. However, investigators often use series connections of two or more heat flux plates (Fuchs and Hadas, 1972) because signals from heat flux plates decrease rapidly when set deeper in the soil. Kimball and Jackson (1975) developed a null alignment to improve the estimate of λ at $Z = Z_1$ (in Eq. [9]); their method avoided using heat flux plates and also assured accounting for sensible and latent heat flux in the soil. Investigators who have used the combination method with the thermal-gradient estimate at depth Z_1, are Kimball et al. (1976a, 1976b) and Kimball and Jackson (1975). Examples of using the combination method with heat flux plates at depth Z_1 are Allmaras et al. (1977), Fuchs and Tanner (1968), and Hanks and Jacobs (1971).

Soil Temperature Models Available

Existing soil temperature models form two general groups: 1) process oriented models which require exact and detailed initial and boundary inputs, and 2) semi or non-process-oriented models which utilize weather station information and soil information at one depth. Several examples of each group will be discussed in relation to type and frequency of input, output accuracy, and sensitivity to soil conditions created or altered by tillage. It is expected that soil temperature models that are sensitive to a range of soil physical conditions will include or account for parameters required for H_g predictions.

PROCESS-ORIENTED MODELS

Hanks et al. (1971) used a numerical approximation for the one-dimensional soil-heat-flow equation. The approximation:

$$[T_j^n - T_j^{n-1}]/\Delta t = [(T_{j-1}^n - T_j^n)D_{j-1/2}^n - (T_j^n - T_{j+1}^n)D_{j+1/2}^n]/\Delta Z^2 \qquad [10]$$

can be solved to predict the temperature of the jth soil depth increment at time increment n. The thermal diffusivity (soil thermal conductivity/soil heat capacity) is given by D. By solving the equation for each soil increment, the soil temperature for each depth and time can be predicted. This method requires the input of initial and boundary conditions and the soil thermal conductivity and heat capacity as a function of depth (Z) and time (t).

Soil temperature was generally predicted within $\pm$ 1 C for a 3-day time period. It was noted that, if soil diffusivity could be estimated within a factor of two, soil temperature could be predicted for a 3-day period within $\pm$ 2 C in a soil slowly drying due to evapotranspiration. Tillage altered soil properties which are treated by this model include soil thermal conductivity and soil heat capacity.

Wierenga et al. (1969) applied the explicit finite difference equation (Richtmeyer and Morton, 1967):

$$[T_j^{n+1} - T_j^n]/\Delta t = [D(T_{j+1}^n - 2T_j^n + T_{j-1}^n)]/\Delta Z^2 \qquad [11]$$

where j is the depth interval and n is the time interval, to approximate the one dimensional heat flow equation.

Thermal diffusivity, D, was estimated from observed temperature variations at several soil depths. With an estimated value of D, known temperature variations with time at 10 and 125-cm depths and a measured initial vertical temperature distribution, soil temperatures were calculated over a 24-hour period for 5-cm soil increments. The procedure was repeated for increasing values of D. An acceptable D for a day was determined when the smallest temperature difference occurred at all depths and times.

Soil temperatures on irrigated plots were predicted within 0.3 C of the observed soil temperature when the 10-cm depth was considered the upper boundary. When the upper boundary was 0 or 5 cm, predicted soil temperature did not agree with the observed soil temperature. Prediction errors were caused by a non-uniform thermal diffusivity in the top 10 cm of soil, which was due to non-uniform soil-water content distribution in this region. On unirrigated plots, no single diffusivity value would yield an acceptable relationship between predicted and observed soil temperature values.

This approach would seem to be tillage sensitive since D is affected by tillage. However, the prediction errors that resulted from non-uniform soil thermal properties with depth indicate that this method may not be sensitive for most tillage-induced soil conditions.

Wierenga and de Wit (1970) developed a simulation model to predict subsoil temperatures from temperature variation at the soil surface and the thermal diffusivity, which is dependent upon depth below the soil surface and soil temperature. A method was developed to predict the thermal conductivity of each soil increment upon the basis of the de Vries (1963) equation:

$$\lambda = \sum_{i=1}^{n} K_i X_i \lambda_i / \sum_{i=1}^{n} K_i X_i \qquad [12]$$

where X_i is the volume fraction of each soil constituent, λ_i is its thermal conductivity, and n is the number of soil constituents. The K_i depends upon the shape and orientation of the soil granules and the ratio between the conductivities of the constituents.

Heat flow through the soil profile was determined by the temperature difference between soil increments, the distance between increment centers, and the average apparent thermal conductivity. The temperature of each soil increment was determined by the quotient of the heat content divided by the volumetric heat capacity and increment thickness.

For a uniform soil with a sinusoidal surface temperature variation, predicted soil temperature values generally were within ± 3 C of the observed soil temperatures. The largest differences occurred near the soil surface. Closer agreement between observed and predicted values was obtained in wet soils than in dry soils.

The inputs required for Wierenga and de Wit's model include initial boundary soil temperature conditions, soil water content, bulk density, and percentage of quartz and organic matter as a function of depth. The soil thermal conductivity and heat capacity are important tillage-influenced components of this model. The model requires testing under non-uniform or layered soil conditions, such as those developed by tillage, before recommendation could be made to utilize this model for predicting tillage effects on soil temperature.

Stated problems with estimating soil temperatures in the shallow soil layers by using the Wierenga and de Wit (1970), Wierenga et al. (1969), and Hanks et al. (1971) process-oriented models suggest a need to more accurately simulate the heat flows due to vapor movement responding to thermal and vapor pressure gradients. An example of such a computation is the simulation of Kimball et al. (1976b) and Hammel[4]. A thin dry layer overlying a wetter layer is always a possibility for tilled soils. In addition to a better accounting of soil temperature, simulations of coupled heat and moisture transfer in the upper 20 cm could provide feedback information to realistically estimate diurnal changes in soil heat flux.

Van Bavel and Hillel (1975, 1976) developed a dynamic numerical procedure to link together the process-oriented simulations of heat movement in the soil and the partition of heat and energy at the soil surface. Soil surface temperature, T_0, is calculated as a factor in predicting evaporation from a bare soil. Their technique utilized simultaneous solution of seven equations with seven unknowns—net radiative flux, evaporation rate, air sensible heat flux, soil sensible heat flux, surface soil temperature, Richardson's number, and the saturation humidity at the surface soil temperature. Of these seven unknowns, only the surface soil temperature could not be stated explicitly. Heat and water (liquid) flows are each coupled at the soil surface. An iterative procedure was thus used at each update to find the proper soil surface temperature. Soil temperature profiles were then estimated (Wierenga and de Wit, 1970) by using these estimates of T_0 as the surface boundary condition. Soil thermal properties with depth were determined by the de Vries (1963) method. Soil temperatures with depth were calculated in a stepwise manner and updated hourly for each soil increment.

Inputs required for this model include solar radiation, air dewpoint temperature, wind speed, initial soil temperature profile, and the surface roughness evaluated by its effect on the aerodynamic roughness parameter, Z_0. No comparisons were made between predicted and measured soil temperature. In their 1975 simulations, evaporation and soil temperatures were simulated for a dry mulch overlying the bare soil; the radiant balance was assumed at the mulch surface, and the heat balance at the base of the mulch. Tillage-sensitive parameters in this model are soil thermal properties, soil surface roughness as related to albedo and aerodynamic properties, and soil hydraulic conductivity. This model requires extensive computer facilities, but perhaps such simulation intensity should be expected as consistent with the multitude of interactions involved with soil heat flux. The authors support their intensive simulation as an action necessary to account for the interacting subsurface and surface factors (including heat flux) that may affect evaporation. Their model is proposed as a more useful approach for estimating evaporation than simulations which use potential evaporation as a forcing function.

Rosema (1975) has developed, for remote-sensing purposes, a simulation similar in most aspects to the van Bavel and Hillel (1976) simulation.

---

[4]Hammel, J. E. 1979. Modeling tillage effects on evaporation and seedzone water content during fallow in eastern Washington. Ph.D. Thesis. Washington State University, Pullman, Wash.

He included a coupled heat and moisture flux throughout the soil profile and a water balance at the surface. Formulations for heat and vapor transfer in the air boundary layer were more complex than those used by van Bavel and Hillel (1975). There were no simulations to test sensitivity to tillage related parameters. Hammel[4] developed a simulation of simultaneous heat and water flow in soil, and applied it to the prediction of soil temperature and water during a drying period in which the surface was either untilled or tilled with a rod weeder. There were separate heat and water flow equations, each with a source term. The source term for the heat flow equation contained the latent heat transfer by isothermal vapor flow, and that for the water flow equation contained thermally induced vapor flow. The soil heat flow equation was bounded at the soil surface by an air temperature estimated from radiative, aerodynamic, and eddy transfer functions. The soil water flow equation was bounded at the surface by an evaporation rate based on air humidity profile linked to the soil water content at the surface. Thus the necessary micrometeorologic measurements were windspeed, incoming radiation, wet and dry bulb, and occasional soil surface temperature. Required parameters indicative of tillage input were roughness length for momentum, roughness parameter for sensible heat exchange, albedo, and longwave emissivity. Soil properties required were the moisture characteristic curve, isothermal water diffusivity, and thermal conductivity. His soil temperature simulations over a 15-day drying period were within 3 C of observed tilled and untilled soil temperatures. Simulated gravimetric seedzone water contents were within 1% of observed soil water contents. He also performed sensitivity tests showing the importance of correctly accounting for the atmospheric boundary layer resistance. These simulations confirmed the importance of soil roughness and plant residue for controlling soil temperature and moisture.

SEMI OR NON-PROCESS-ORIENTED SOIL TEMPERATURE MODELS

Hasfurther and Burman (1974) used ambient air temperature to predict average daily soil temperatures of different depths. Fourier transform techniques were utilized on several years' data at a given location to develop a linear equation for predicting average daily soil temperature at the 5-cm depth for that location. The root zone temperature was predicted by the one-dimensional heat flow equation as presented by Wierenga et al. (1969). Boundary conditions for the root zone temperature predictions were based on the periodic soil temperature variation function described by van Wijk (1963).

For a crested wheatgrass-covered soil, their 10-day straight running average predicted soil temperatures generally were within 3 C. However, use of this method for modeling tillage effects on soil temperature appears limited. Van Wijk's (1963) method for estimating surface boundary conditions assumes homogeneous and isotropic soil conditions, which generally do not exist for tilled soil conditions. Extensive calibration also would be needed to predict soil temperatures resulting from various tillage-induced conditions for different soils and/or different locations.

Neild (1971) developed regression equations that contained relations between average weekly air temperatures (and week numbers beginning with 1 March) and the weekly average soil temperature at soil depths of < 2.5, 5, 10, and 20 cm under bare or sod covered soils. In the spring and fall, r^2 values were 0.85 or greater but were lower during the summer weeks. Standard errors ranged from 1.06 to 2.59 C. Calibration for each site and tillage operation would be needed to yield average weekly soil temperatures in using this method. This approach is not recommended for modeling soil temperature responses to tillage.

Gupta et al. (1981) estimated hourly soil temperatures by depth. Inputs needed for this model include hourly air temperature at the 2-m height, daily maximum and minimum soil temperatures, initial soil temperature with depth, and soil thermal diffusivity which may be estimated from soil mineral composition, organic matter percentage, bulk density, and soil water content (de Vries, 1963). The model is divided into two sections: 1) estimation of the upper boundary temperature (UBT) and 2) prediction of the root zone temperature upon the basis of UBT and soil thermal properties. The UBT $(T_{o,t})$ is estimated by the equation:

$$T_{o,t} = T_{1,t} + A \sin \left[(2 \, \pi/P) \, t \right] \qquad [13]$$

where $T_{1,t}$ is the air temperature at the 2-m height at time t; A is the difference between the maximum air and maximum surface soil temperature (the difference between the maximum air and maximum soil temperature is used for the daylight period, and the difference between the minimum air and minimum soil temperatures is used for the night period); P is twice the time period that elapses from the time when both air and surface soil temperature are equal to the next time when both air and surface soil temperature are equal; and t is equal to the difference in time between the present time and the last time when air temperature and surface soil temperature were equal. Relationships between A and percentage surface cover were obtained from field data. This gave a means of accounting for residue-cover effects on $T_{0,t}$ (or the UBT).

The Hanks et al. (1971) numerical approximation, utilizing the UBT calculated by Gupta et al. (1981), served as the root zone temperature model.

Soil temperature predictions were made hourly for a soil: 1) without a crop and covered (0 to 99% coverage) with oatstraw mulch treatments and 2) with a corn crop and either a bare soil surface or a 50% oatstraw-covered surface. Because of the field-measured relationships between A and percentage surface cover, their model gave a means of accounting for residue-cover effects on $T_{0,t}$ (or the UBT). Predicted hourly UBT generally varied within ± 8.9 C and − 3.2 to 4.4 C of the measured UBT for bare and 50% residue-covered soil surfaces, respectively. The predicted root zone temperatures at 5, 10, and 30-cm depths were within 2 to 10 C of the measured values.

With further calibration, it may be possible to apply this method to tillage research. The model accounts for mulch effects on soil temperature by using A as input sensitive to residue cover. Evaluating soil thermal dif-

fusivity, in the equation for predicting root zone temperature, as a tillage
sensitive variable would further enhance use of this model for tillage-
created soil conditions. Because matted and standing residues may cause
energy and heat exchanges not fully accounted for by albedo and aerodynamic parameters, process-oriented models may be calibrated by using
measures of A as done by Gupta et al. (1981).

Cruse et al. (1980) predicted tilled soil temperatures on the basis of
the concept that temperature differences between a bare and residue-
covered soil can be calculated as a function of absorbed solar radiation
(Van Doren and Allmaras, 1978) and soil thermal inertia (Lettau, 1951):

$$T_1 = T_2 - (R_s - R_s') (\alpha_1 - \alpha_2) (K) (I_1/I_2) \qquad [14]$$

where
  T_1 = average daily 5-cm soil temperature of a residue treated and/or
       tilled soil (C)
  T_2 = average daily 5-cm soil temperature on a bare untilled soil (C)
  R_s = measured daily incoming solar radiation (W/cm^2)
  R_s' = daily incoming solar radiation at the winter solstice when it is
       assumed $T_2 = T_1$
  α_1 = reflection coefficient of the tilled soil surface
  α_2 = reflection coefficient of the flat bare untilled soil
  K = constant (C cm^2/W)
  I_1 = soil thermal inertia of the residue treated and/or tilled soil (W
       C cm^{-2} sec$^{1/2}$)
  I_2 = soil thermal inertia of the bare untilled soil (W C cm^{-2} sec$^{1/2}$)
Subroutines were developed to estimate changes in solar radiation
absorption with time due to residue decomposition, soil water content,
and soil surface roughness.

Soil and residue properties needed for this soil temperature model include: soil water content at the wilting point, soil bulk density, soil
random roughness, initial residue dry weight, residue reflection coefficient at model initiation, soil reflection coefficient (for the smooth, dry
condition), and percentage of quartz, clay minerals and organic matter in
the soil. Daily soil and climatic inputs include maximum and minimum
air temperature, incoming solar radiation, wind travel, soil water content
at the 5-cm depth, and stage of evaporation.

Predicted soil temperatures generally were within ± 2 C for four
tillage-induced soil conditions. Errors in these predictions seemed to be
treatment and weather dependent. A 4.65-cm rainfall during the test
period altered color and density of surface residue, and reduced roughness
of the soil surface. These processes were not accounted for in the model;
consequently, error patterns in the soil temperature predictions were
changed by this event. The effect of the rainfall and associated soil surface
changes on the prediction errors indicate soil heat flux and/or heat distribution within the soil were affected by these surface changes and that
these changes should be considered as important factors in future modeling efforts.

Simulation Needs Applied to Tillage

Tillage-related factors affecting soil heat flux include: 1) surface radiation reflection, absorption, and emission characteristics; 2) aerodynamic roughness; and 3) surface and subsurface soil thermal properties. Those models discussed herein, with the exception of Neild (1971), include some aspect of soil thermal properties in their soil temperature predictions. Whenever soil thermal properties are needed, most models utilize de Vries (1963) methods for these estimations based on fundamental soil properties. This approach, while comparatively simple and reasonably suitable for most systems to describe flow by conduction and vapor flux, may significantly underestimate effective thermal conductivity in tilled conditions where a significant amount of convection occurs in the soil (Allmaras et al., 1977). Some guidelines are needed concerning the conditions promoting convective enhancement and expected magnitudes of convective enhancement. Tentative guidance on this problem may be obtained from Hadas (1977), Allmaras et al. (1977), Kimball (1973), and Scotter and Raats (1969).

The effect of soil surface conditions on soil radiation absorption and/or emission characteristics, as it directly affects soil temperature, was considered only by Cruse et al. (1980). Linden (1979)[3] has theoretically modeled the relationship between surface properties and radiation absorption. Much more field testing is needed before general use of these approaches can be recommended.

Aerodynamic conditions were included in the simulations of van Bavel and Hillel (1975, 1976) without evaluation for tillage. Hammel[4] used roughness lengths, for momentum, of 0.1 and 1 cm, respectively, for conventionally rod weeded and non-tilled Walla Walla silt loam. These corresponded to roughness element heights of 1 and 10 cm. Different tillage-induced random (or oriented) roughness of the surface most certainly affects the transfer of heat and vapor in the atmospheric boundary layer up to 1 or 2 m. The amount of residue and its physical arrangement, such as matted or standing, must also influence transfers in this boundary layer. Van Doren and Allmaras (1978), attempted to depict the range of aerodynamic parameters expected from tillage roughness and residue-covered surfaces. Aase and Siddoway (1980) found significant differences in estimates of these parameters (roughness length, friction velocity, and zero-plane displacement) for different wheat stubble heights. Webb (1965) and Monteith (1973) give examples of comprehensive analyses of aerial microclimate. Generally these micrometeorological oriented studies contain no reference to tillage related effects on aerodynamics or eddy diffusion. Models such as those by van Bavel and Hillel (1976) and Hammel[4] might be used to test sensitivity to suggested and measured variations in aerodynamic roughness length, zero plane displacement, and resistance to parameters for heat and vapor flux. Here again field estimates must be made to verify projections based on model sensitivity.

Model Input, Output, and Use

Time intervals for soil temperature predictions are dependent upon the intended use of the information. Germination or root and shoot growth are generally nonlinear functions of temperature. Hourly temperature predictions (Gupta et al., 1981; Blacklow, 1973) therefore are needed. Fluctuating temperatures, such as those that occur diurnally, also may produce plant growth responses dissimilar to those resulting from a constant temperature equal to the average over the cycle (Walker, 1970; Blacklow, 1973). Voorhees et al. (1981) concluded that large differences between observed and predicted plant growth response curves often have resulted from using constant or average daily soil temperatures rather than temperatures at shorter time intervals. These nonlinear and fluctuating soil temperature influences are critical factors in model output because temperature differences as small as 1 C can have a significant effect on plant growth (Walker, 1969). Indirect factors, such as nutrient transformation (Stanford et al., 1973) and transpiration (Barlow et al., 1977), which also are temperature sensitive, would be better understood if cyclic soil temperatures were used instead of a constant temperature. Certain plant development parameters, such as leaf rate per day, are much less sensitive to diurnal temperature fluctuations than is carbon assimilation in plants (J. T. Ritchie, personal communication. Temple, Tex.).

In some cases, one may be interested in modeling the end point of a relatively long-term transient of soil temperature (and also evaporation). In other words, output from the soil temperature model would not be simultaneous input to a plant growth function. An example would be control of soil surface roughness by fall tillage to control soil temperature during early growth of spring-seeded corn (Allmaras et al., 1972).

Frequency of measured inputs also is a critical element in model development because tilled layers and even those untilled change with time and therefore influence the processes involved with soil heating. One rainfall event and associated soil surface alterations can significantly affect the soil heating process and/or soil temperaure development (Cruse et al., 1980). This observation clearly indicates that methods must be developed and tested to predict changes in residue cover, residue color, surface roughness, soil bulk density, and soil water content as functions of climate, time, and soil.

Many factors affect the desired complexity of process orientation in a soil temperature model. Input and output frequency are involved as discussed earlier; complexity and availability of input parameters also must be considered. The first objective of a soil temperature model should be identification of the important processes involved in soil heating. Processes and soil heating sensitivity to tillage parameters would most certainly be desired. The process orientation ought to be complete enough to accommodate micrometeorological input. With inclusion of micrometeorological input, there is then a good opportunity to use soil temperature modeling in soil management. Repeated simulations can be made, each utilizing one set of micrometeorological inputs obtained from 1 year of weather observation at a site. It is necessary that the tillage parameters are dynamically adjusted to account for their natural change caused by micrometeorologic events, biological activity, and tillage operations. These simulations provide a probabilistic approach to soil management objectives (Ritchie, 1981; Haan, 1979). Specific scenarios of weather, such as early cold and dry spells followed by wet and warm spells, could give additional insights into tillage management. Tillage-related parameters also can be varied as typical of conservation or conventional alternatives to understand climatic impacts and other constraints on acceptance of conservation tillage.

This review indicates that existing semi and non-process-oriented models are not likely to further improve accuracy of modeled soil temperature or to give a realistic accounting of tillage effects on soil temperature. Future tillage sensitive soil temperature models will require more process orientation to simulate the complex interactions occurring at the soil-air interface and also must account for water movement within and out of the soil. Advances will probably occur in three phases. The first of these, and one which has had significant progress, involves identification of important soil and climatic factors involved in the soil heating process. The second phase, and presently an active research area, involves explaining the effect of the important soil and climatic factors on soil heating. Also during this phase, models to predict the surface and plow layer physical condition with time will or should be developed. The final phase will be that of combining the knowledge gained in Phase 1 and 2 into a working model for predicting soil temperatures under field conditions. In this final phase, model simplification may occur.

LITERATURE CITED

1. Aase, J. K., and F. H. Siddoway. 1980. Stubble height effects on seasonal microclimate, water balance and plant development of no-till winter wheat. Agric. Meteorol. 21:1–20.

2. Allmaras, R. R., E. A. Hallauer, W. W. Nelson, and S. D. Evans. 1977. Surface energy balance and soil thermal property modifications by tillage-induced soil structure. Minnesota Agric. Exp. Stn. Tech. Bull. 306. 41 p.

3. ————, W. W. Nelson, and E. A. Hallauer. 1972. Fall versus spring plowing and related soil heat balance in the western cornbelt. Minnesota Agric. Exp. Stn. Tech. Bull. 283. 21 p.

4. Barlow, E. W. R., L. Boersma, and J. L. Young. 1977. Photosynthesis, transpiration and leaf elongation in corn seedlings at suboptimal soil temperatures. Agron. J. 69:95–100.

5. Bertrand, A. R. 1980. Overdrawing the nations research accounts. J. Soil Water Conserv. 35:109–111.

6. Blacklow, W. M. 1973. Simulation model to predict germination and emergence of corn (Zea mays) in an environment of changing temperature. Crop Sci. 13:604–608.

7. Brutsaert, W. 1975. On a derivable formula for long-wave radiation from clear skies. Water Resour. Res. 11:742–744.

8. Cruse, R. M., D. R. Linden, J. K. Radke, W. E. Larson, and K. Larntz. 1980. A model to predict tillage effects on soil temperature. Soil Sci. Soc. Am. J. 44:378–383.

9. de Vries, D. A. 1963. Thermal properties of soils. p. 210–235. *In* W. R. van Wijk (ed.) Physics of the plant environment. John Wiley and Sons, Inc., New York.

10. Fuchs, M., and A. Hadas. 1972. The heat flux density in a nonhomogeneous bare loessial soil. Boundary Layer Meterol. 3:191–200.

11. ————, and C. B. Tanner. 1968. Calibration and field test of soil heat flux plates. Soil Sci. Soc. Am. Proc. 32:326–328.

12. Gupta, S. C., J. K. Radke, and W. E. Larson. 1981. Predicting temperature of bare and residue covered soils with and without a corn crop. Soil Sci. Soc. Am. J. 45:405–412.

13. Haan, C. T. 1979. Risk analysis in environmental modification. p. 30–51. *In* B. J. Barfield and J. F. Gerber (ed.) Modification of the aerial environment of plants. ASAE Monograph 2. Am. Soc. Agric. Eng., St. Joseph, Mich.

14. Hadas, A. 1977. Evaluation of theoretically predicted thermal conductivities of soils under field and laboratory conditions. Soil Sci. Soc. Am. J. 41:460–466.

15. Hanks, R. J., D. D. Austin, and W. T. Ondrechen. 1971. Soil temperature estimation by a numerical method. Soil Sci. Soc. Am. Proc. 35:665–667.

16. ————, and H. S. Jacobs. 1971. Comparison of the calormetric and flux meter measurements of soil heat flux. Soil Sci. Soc. Am. Proc. 25:671–674.

17. ————, and C. B. Tanner. 1972. Calorimetric and flux meter measurements of soil heat flow. Soil Sci. Soc. Am. Proc. 36:537–538.

18. Hasfurther, V. R., and R. D. Burman. 1974. Soil temperature modeling using air temperature as a driving mechanism. Trans. ASAE 17:78–81.

19. Idso, S. B., J. K. Aase, and R. D. Jackson. 1975. Net radiation-soil heat flux relations as influenced by soil water content variations. Boundary Layer Meteorol. 9:13–122.

20. ————, and R. D. Jackson. 1969. Thermal radiation from the atmosphere. J. Geophys. Res. 74:5397–5403.

21. Jensen, M. E. (ed.). 1974. Consumptive use of water and irrigation water requirements. Am. Soc. Civil Engineers, New York, N.Y.

22. Kimball, B. A. 1973. Water vapor movement through mulches under field conditions. Soil Sci. Soc. Am. Proc. 37:813–818.

23. ————, and R. D. Jackson. 1975. Soil heat flux determination: a null alignment method. Agric. Meteorol. 15:1–9.

24. ————, ————, F. S. Nakayama, S. B. Idso, and R. J. Reginato. 1976a. Soil heat flux determination: temperature gradient method with computed thermal conductivities. Soil Sci. Soc. Am. Proc. 40:25–28.

25. ————, ————, R. J. Reginato, R. S. Nakayama, and S. B. Idso. 1976b. Comparison of field measured and calculated soil heat fluxes. Soil Sci. Soc. Am. Proc. 40:18–25.

26. Larson, W. E. 1964. Soil parameters for evaluating tillage needs and operations. Soil Sci. Soc. Am. Proc. 38:118–122.

27. Lettau, H. H. 1951. Theory of surface-temperature and heat transfer oscillations near a level ground surface. Trans. Am. Geophys. Union. 32:189–200.

28. Linden, D. R. 1981. Predicting tillage effects on evaporation from soil. p. 117–132. *In* Predicting tillage effects in soil physical properties and processes. ASA, SSSA Spec. Pub. no. 44, Madison, Wis.

29. List, R. J. 1968. Smithsonian meteorological tables. 6th revised ed. Smithsonian Inst. Press.

30. Monteith, J. L. 1973. Principles of environmental physics. Am. Elsevier Publ. Co., Inc., New York, N.Y.

31. Neild, R. E. 1971. Growing season air-soil temperature relationships at Lincoln, Nebraska. Nebr. Agric. Exp. Stn. Res. Bull. 242.

32. Penman, H. L. 1948. Natural evaporation from open water, bare soil and grass. Proc. R. Soc. London Ser. A193:120–145.

33. Phillip, J. R. 1961. The theory of heat flux meters. J. Geophys. Res. 66:671–679.

34. Price, J. C. 1980. Potential of remotely sensed thermal infrared data to infer surface soil moisture and evaporation. Water Resour. Res. 16:787–795.

35. Richtmeyer, R. D., and K. W. Morton. 1967. Different methods for initial-value problems. 2nd ed. Interscience Publ., New York. 405 p.

36. Ritchie, J. T. 1981. Water management and water efficiencies for American agriculture. p. 15–42. In Research and its application to crop production in the 1980's. Proc. of a Weather and Agriculture Symp. 1–2 Oct. 1979. Kansas City, Mo. Univ. of Missouri, Columbia.

37. Rosemā, Andries. 1975. Simulation of the thermal behavior of bare soils for remote sensing purposes. p. 109–123. In D. A. de Vries and N. H. Afgan (ed.) Heat and mass transfer in the biosphere; Part I. Transfer processes in the plant environment. John Wiley and Sons, Inc., N.Y.

38. Scotter, D. R., and P. A. C. Raats. 1969. Dispersion of water vapor in soil due to air turbulence. Soil Sci. 108:170–176.

39. Sears, F. W., and M. W. Zemansky. 1970. University Physics. Addison-Wesley Publishing Co., Inc., Reading, Mass.

40. Stanford, G., M. H. Frere, and D. H. Schaninger. 1973. Temperature coefficient of soil nitrogen mineralization. Soil Sci. 115:321–323.

41. Tanner, C. B. 1968. Evaporation of water from plants and soil. p. 73–106. In T. T. Kozlowski (ed.) Water deficitys and plant growth. I Development, control, and measurement. Academic Press, New York.

42. van Bavel, C. H. M., and D. I. Hillel. 1975. A simulation study of soil heat and moisture dynamics as affected by a dry mulch. p. 815–821. In Proc. of 1975 Summer Computer Simulation Conference, San Francisco, Calif. Simulation Councils Inc., La Jolla, Calif.

43. ————, and ————. 1976. Calculating potential and actual evaporation from a bare soil surface by simulation of concurrent flow of water and heat. Agric. Meteorol. 17:453–476.

44. Van Doren, D. M., Jr., and R. R. Allmaras. 1978. Effect of residue management practices on the soil physical environment, microclimate, and plant growth. p. 49–83. In W. R. Oschwald (ed.) Crop residue management systems. Am. Soc. Agron. Spec. Publ. 31.

45. van Wijk, W. R. 1963. Physics of plant environment. John Wiley and Sons, Inc., New York.

46. Voorhees, W. B., R. R. Allmaras, and C. E. Johnson. 1981. Root zone modification for alleviating temperature stress. In Modifying the plant root environment. Am. Soc. Agron. Monogr. (In preparation).

47. Walker, J. M. 1969. One degree increments in soil temperature affect maize seedling behavior. Soil Sci. Soc. Am. Proc. 33:729–736.

48. ————. 1970. Effects of alternating versus constant soil temperatures on maize seedling growth. Soil Sci. Soc. Am. Proc. 34:889–892.

49. Webb, E. K. 1965. Aerial microclimate. Meteorol. Monogr. 6:27–58.

50. Wierenga, P. J., and C. T. de Wit. 1970. Simulation of heat flow in soils. Soil Sci. Soc. Am. Proc. 34:845–848.

51. ————, D. R. Nielsen, and R. M. Hagan. 1969. Thermal properties of a soil based upon field and laboratory measurements. Soil Sci. Soc. Am. Proc. 33:354–360.

Chapter 10

Modeling Soil Mechanical Behavior During Tillage[1]

S. C. GUPTA AND W. E. LARSON[2]

ABSTRACT

This manuscript discusses models for predicting soil mechanical behavior during tillage operations. Soil mechanical behavior is discussed in terms of (a) soil breakup and (b) soil compaction.

The soil breakup model is based on the assumptions that (a) the optimum soil physical (water storage, heat flow, and gaseous diffusion) or biological (microbial activity, seed germination, and seedling growth) processes are associated with bulk density of any given soil, (b) the packing bulk density of medium to fine-textured agricultural soils is related to the aggregate size distribution and the aggregate characteristics, and (c) the energy required to create different aggregate size distributions varies with the water content and the type of soil. Thus, if the optimum bulk density is known, then the procedure suggested in the soil breakup model can be used to select tillage implements for a soil at a given water content that would provide aggregate size distribution optimum for a soil process.

The soil compaction model is based on the empirically derived relationship between bulk density and applied stress at different water contents. Inputs to the model are particle size analysis, the type of clay, the water contents at the time of vehicular traffic, traffic load, and the dimensions of the contact area between the tire and the soil surface. Output from the model is the change in the porosity of soil as influenced by water content and applied load. Porosity changes with water content and applied stress are then used to show a range of water contents and vehicular stresses that do not cause detrimental soil compaction. Criteria used for detrimental soil compaction are (a) air-filled porosities critical for gaseous diffusion, (b) stresses critical for shearing soil aggregates, and (c) soil resistance critical for root growth. Predicted stress and bulk density profile after a passage of vehicular traffic is also shown.

[1] Contribution from the Soil and Water Management Research Unit, North Central Region, USDA-ARS, St. Paul, MN, in cooperation with the Minnesota Agric. Exp. Stn., Paper No. 11,378, Scientific Journal Series.

[2] Soil scientists, USDA-ARS, Univ. of Minn., St. Paul, MN 55108.

INTRODUCTION

A major reason for tilling agricultural soils is to create soil physical conditions that are conducive to good seed germination. Generally, this means desirable temperatures and water and aeration conditions of the seed zone for the given plant species. Soil physical conditions, in turn, affect the microbial activity, root growth, and other biological processes in the soil.

Other reasons for soil tillage are to (a) incorporate crop residues and fertilizers, (b) control weeds, (c) minimize soil erosion by water and wind, (d) increase rainfall intake and storage, (e) encourage warming and drying of the seed zone, and (f) minimize the yield-reducing effects of plant pests. The integrated effect of all these factors is reflected in crop yields.

During tillage, a part of the soil is broken-up into various size clods by the implements and a part is compacted by the traffic. Depending upon the soil type, water content at the time of tillage, and stresses exerted by the tillage implements and equipment, soils are affected differently by the breakup and compaction processes. The aim of this manuscript is to present models for predicting soil mechanical behavior during tillage operations. Soil mechanical behavior in this paper is defined in terms of (a) soil breakup and (b) soil compaction.

Soil Breakup

Allmaras et al. (1965) showed that the geometric mean diameter of aggregates in the row zone varies with the type of tillage and the type of soil. Both the tillage implement and the size of the cut by the implement are also important factors in the resultant mean weight diameter (MWD) of the clods for a consolidated cohesive soil (Gill and McCreery, 1960). The mean weight diameter of aggregates was linearly related to the $\log_{10}$ of the input energy in the drop shatter test. They used the drop shatter energy and the draft measurements in the field to evaluate the efficiency of different tillage tools. Farrell et al. (1967) indicated that the cumulative drop height in the drop shatter test is a good measure of the soil fragmentation, because of its constant proportionality with the tensile strain energy. Bateman et al. (1965) compared two methods of measuring energy for breaking up soil at different degrees of compaction and soil moisture contents. Soil considered in this study was Drummer silty clay loam—an Illinois soil that is difficult to pulverize. They concluded that (a) the impact method of pulverizing the soil requires more energy than the slow loading method; (b) drying low bulk density (1.16 g cm^{-3}) soil has little effect on the energy requirement by either loading method; (c) drying high bulk density (1.40 to 1.56 g cm^{-3}) soils significantly increases the energy requirement for pulverization by both methods; (d) initially compacting the soil increases the energy requirements more for the finer degree of pulverization (MWD $= 1.3$ cm) than for coarse pulverization (MWD $= 10.2$ cm).

The degree of soil breakup that is optimum for plant growth depends upon the seed size, the crop type, and the soil and weather conditions. For example, Larson and Swan (1970) suggested an average aggregate diameter of 6 mm in the row zone of wet soils for corn. They indicated that with most planters, this would produce good packing over the seed at a water content favorable for germination. For dry soil, they suggested use of furrow openers or listing to place the seed in moist soil and a bed of 2 to 6-mm aggregates at the soil surface to slow evaporation.

Soil Compaction

Harris (1971) discussed several models that describe soil compaction. All these models are empirically derived relationships between the bulk density or porosity or void ratio and the applied loads on finite soil columns. These relationships are either exponential or logarithmic functions. From laboratory confined compression tests, Larson et al. (1980) showed that (a) at a given water content, compression curves (bulk density vs. log applied stress) determined on agricultural soils are linear over the range of stress from about 1 to 10 kg cm^{-2} and (b) compression curves at different water contents for a given soil are approximately parallel over the range of initial pore water potential from -0.05 to -0.7 bars. The relationship that described the linear portion of the compression curve is:

$$\varrho = [\varrho_K + \Delta_T (S_1 - S_K)] + C \log (\sigma_a/\sigma_K) \qquad [1]$$

where

ϱ = compacted bulk density, g cm^{-3}, corresponding to an applied stress, σ_a.

ϱ_K = bulk density, g cm^{-3}, at a known stress $\sigma_K = 1$ kg cm^{-2}.

Δ_T = slope of the bulk density vs. degree of water saturation curve at σ_K. This is the same as S_T in Larson et al. (1980).

S_1 = desired degree of saturation at σ_K, %.

S_K = degree of saturation corresponding to ϱ_K and σ_K, %.

C = compression index (slope of the linear portion of the compression curve).

Equation [1] is similar to the models discussed by Harris (1971).

Soehne (1953) described a procedure for calculating stress distribution in the ground after a passage of vehicular traffic. This procedure is based on the Boussinesq equation as modified (Frohlich, 1934) by the introduction of a concentration factor to represent different kinds and conditions of soil. Equations that describe various stresses on an element of soil (Fig. 1a) as a result of concentrated point load, P, at point 0 in xy-plane are as follows:

$$\sigma_z = \frac{\upsilon P}{2\pi r^2} \cos^\upsilon \beta \qquad [2]$$

$$\sigma_h = \frac{\upsilon P}{2\pi r^2} \cos^{(\upsilon - 2)} \beta \sin^2 \beta \qquad [3]$$

$$\Upsilon = \frac{\upsilon P}{2\pi r^2} \cos^{(\upsilon - 1)} \beta \sin \beta \qquad [4]$$

where

σ_z = the vertical normal stress, kg cm^{-2}

σ_h = the horizontal normal stress in a radial direction

Υ = the shearing stress belonging to σ_z and σ_h

r = the radial distance in the soil profile from point 0

β = the angle bisected by a vertical line from point 0 with a line to the center of gravity of the volume element under question

v = the concentration factor

Soehne (1953) assumed that the contact area between tires and ground was elliptical in shape (Fig. 1b). The procedure for calculating main stress in the profile involved dividing the elliptical load surface into 25 load elements in whose center of gravity, a fraction of the surface load was applied as a concentrated load. Main stress at a point in the soil profile was then calculated from the summation of stresses (Eq. [2], [3], and [4]) from 25 load elements. Readers should refer to Soehne (1953) article for equations and other details of this procedure. In Eq. [2], [3], and [4], the greater the value of v, the more concentrated are the stresses towards the load axis. Soehne (1953) suggested v equal to 4, 5, and 6 for hard, firm, and soft conditions of the soil, respectively. Soehne's description of firmness applies to a combination of both bulk density and water status of the soil.

Using a combination of two different size tires and two different loads, Taylor et al. (1978) concluded that Frohlich's equation (Eq. [2]) approximately described the soil pressure distribution in both a sandy soil and a clay soil with uniform density profiles. However, when a compacted layer was introduced in the density profile, this equation was no longer suitable. Blackwell and Soane (1981) also showed that after passage by an agricultural vehicle, Frohlich's equation required a different concentration factor with soil depth in order to match predicted bulk density profiles with measured values.

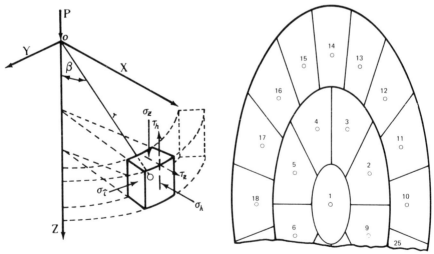

Fig. 1. (a) Stresses on a soil element caused by a point load, P. (b) A surface load broken up into 25 load elements. Open circles represent the center of gravity with a pressure distribution according to a parabola of 4th potency (Soehne, 1953).

Effects of soil compaction can be described in three ways: (1) decrease in air-filled porosity, (2) degradation of soil aggregates, and (3) increase in the friction of soil to roots. Depending upon the degree of soil compaction, these effects may or may not be harmful to plants. The following discussion deals with the critical values of the variables that define detrimental soil compaction.

Vomocil and Flocker (1961) concluded that air-filled porosity of less than 10% when water content is at field capacity can be expected to restrict root growth and thus plant development. Restricted root growth could be due to inadequate oxygen supply or limiting rate of water supply to roots or excessive mechanical impedance of soil to roots. When air-filled porosity is reduced beyond a critical limit, diffusion of oxygen and CO_2 between the soil and the atmosphere cannot be maintained at a rate suitable for biological activities. Grable (1971) showed that for medium rate of respiration, oxygen percentage in the soil approached zero when air-filled porosity approached 12%. From the results of several researchers, Wesseling and van Wijk (1957) concluded that the diffusion of CO_2 in the soil was nearly zero at air-filled porosity of less than 10%.

Larson and Gupta (1980) showed that during compression of unsaturated agricultural soils the pore water potential decreased to a minimum and then increased as the applied mechanical stress increased. Apparently, as the applied stress increased beyond the point of minimum pore water potential (U_m), soil aggregates are sheared and their identity destroyed. They defined σ_a corresponding to the point of minimum pore water potential as the critical stress (σ_c):

$$\log \sigma_c = \sigma_n \log \sigma_s \qquad [5]$$

where
σ_n = the normalized log of stress at U_m and
σ_s = the stress when the soil is saturated.

Since degradation of soil structure can decrease plant growth and lower water infiltration, Larson and Gupta (1980) suggested that the maximum stress applied to a soil during cultivation or vehicular traffic should be less than the critical stress.

Taylor and Gardner (1963) and Taylor and Bruce (1968) have shown that resistance to a steel probe (penetrometer) inserted in the soil is empirically related to root growth. In turn, penetration resistance is related to the bulk density and the water content of the soil. These authors also showed that the penetration of a cotton tap root was nearly zero when the resistance of soil to a cylinderical steel probe (0.5 cm diam) was above 20 kg cm^{-2}.

Voorhees et al. (1975) measured the primary root elongation rates of pea seedlings in a sandy and a clay soil at -0.1, -0.33, and -1.0 bar matric potentials. They showed that the primary root elongation rate decreased (0.09 to 0.02 cm hour^{-1}) with an increase (0.04 to 15 kg cm^{-2}) in the normal point resistance, σ_N. Normal point resistance is related to the total point resistance, σ_P, by the relationship:

$$\sigma_P = \sigma_N (1 + \tan \phi' \cot \alpha) \qquad [6]$$

where

ϕ' = coefficient of soil metal friction

α = included semiangle of the conical probe.

Normal point resistance of 15 kg cm^{-2} corresponded to a total point resistance of approximately 20 kg cm^{-2} in a clay soil with $\alpha = 30°$. This value of critical penetration resistance is similar to the one suggested by Taylor and Gardner (1963) and Taylor and Bruce (1968). However, the value of the limiting resistance would vary with the type of roots and the type of penetrometer.

Reviews of soil compaction, its causes and effects, have been given by Barnes et al. (1971) and Chancellor (1976). Readers should refer to these references for further details.

The main objective of this paper is to present models that describe the process of soil breakup and soil compaction during tillage operations. Because we lacked enough theoretical or empirical relationships for the construction of these models, we devised several experiments to develop new relationships and to test existing hypotheses from the literature. Following is the list of experimental objectives that were used to build soil breakup and soil compaction models:

1. To develop the relationship between the energy required to break up soil clods and the geometric mean diameter of the shattered aggregates.
2. To predict the bulk density of medium and fine-textured agricultural soils from the aggregate characteristics and aggregate size distribution.
3. To predict from particle size distribution of soils, changes in the total porosity of the soil as influenced by the water content and the amounts of applied mechanical stress during uniaxial compression tests.
4. To predict a range of soil water content and vehicular stress that does not create detrimental soil compaction at the time of tillage.
5. To calibrate Frohlich's equations using the laboratory measured bulk density profile in a soil column after an application of load on an elliptical plexiglass plate simulating the shape of the contact area between the tire and the soil.
6. To predict the effect of agricultural traffic on the soil bulk density profile using particle size analysis.

EXPERIMENTAL PROCEDURES

Experimental procedures are discussed in the order of the experimental objectives given in the last section. Most of the discussion in this manuscript is centered around two soils: Webster clay loam (Typic Haplaquoll) and Nicollet clay loam (Aquic Hapludoll).

Experiment 1

A roughly spherical clod (weighing 300 to 400 g) of undisturbed Webster clay loam at field water content was dropped from a height of 1 m onto a lacquered hardboard surface. Fragments from the shattered clods were sieved through a nest of sieves. This process was then repeated twice with the sieved fragments, thus giving a cumulative drop height of 300 cm. Sieved fragments were recombined before each drop. The sieve classes include > 25.4, 25.4 to 12.7, 12.7 to 5.66, 5.66 to 3.36, 3.36 to 2.0, 2.0 to 1.0, and < 1.0 mm. Geometric mean diameter (GMD) was calculated using the lower diameter (d) limit of each sieve class in the following relationship:

$$\text{GMD (mm)} = \text{antilog} \left(\sum_{i=1}^{N} W_i \log d_i / \sum_{i=1}^{N} W_i \right) \qquad [7]$$

where
W_i = the weight of aggregates in a sieve class i, g.
N = the number of fractions.
Cumulative drop height represents an equivalent energy applied to break up the soil. A relationship was developed between the cumulative drop height and the geometric mean diameter of the shattered aggregates. A similar procedure of shattering soil clods has been discussed by Marshall and Quirk (1950), Gill and McCreery (1960), and Farrell et al. (1967). The drop shatter test was repeated on other clods at water contents less than and greater than the field water content. These water contents were achieved by spraying clods with an atomizer or air-drying. Clods were then stored for several days to equilibrate at each water content before running the drop shatter test.

Experiment 2

A surface (0 to 20 cm) sample of Webster clay loam which has been disturbed by cultivation was air-dried and passed through a series of sieves (Table 1). Aggregates in each sieve class were weighed and geometric mean diameter calculated using Eq. [7]. Aggregates in the same weight proportion as the original sample were then mixed in a plastic bag such that the final quantity of mixture was 4 kg. The mixture of aggregates was then poured into a straight wall plastic container (17.34 cm in diameter and 19 cm in length), and the air-dry bulk density of mixture measured after packing. Packing was done by dropping the container of loose material 50 times from a height of 10 to 15 cm (Gupta and Larson, 1979a). The aggregate mixture was then poured into another container and packed again. This process was repeated 10 times to determine an average bulk density. This procedure was repeated to determine bulk density on six other aggregate size distributions which were created after sieving the contents of the container and by mixing additional amounts of some aggregate fractions. Each of the seven aggregate size distributions represented a dominance by certain aggregate size fractions.

Bulk density of the aggregate mixture was predicted from a packing model (Gupta and Larson, 1979a). Inputs (Table 1) to the packing model were: 1) aggregate size distribution; 2) bulk density of each aggregate fraction, ϱ_b; and 3) bulk density of individual aggregate, ϱ_a. Bulk density

Table 1. Packing densities and geometric mean diameter of various aggregate size distributions of dry Webster clay loam. Boxes indicate the dominating aggregate fractions in a distribution.

Diameter (mm)	ϱ_b	ϱ_a	Distributions number						
			1	2	3	4	5	6	7
			Percent aggregate by weight						
50.8 –25.4	0.93	1.90	40.20	5.01	5.04	5.07	5.25	4.92	5.42
25.4 –12.7	0.97	1.87	30.00	9.35	4.80	4.94	4.96	17.60	4.84
12.7 – 5.66	0.95	1.87	11.30	43.80	5.36	5.37	5.52	6.66	23.20
5.66 – 3.36	0.96	1.89	3.31	10.70	5.27	4.49	4.76	4.49	7.61
3.36 – 2.0	0.97	1.86	2.53	6.58	21.70	4.80	5.12	4.86	5.90
2.0 – 1.0	0.99	1.87	3.90	7.60	19.30	9.81	6.44	5.41	7.65
1.0 – 0.5	1.09	1.90	3.57	6.40	13.90	26.80	6.97	6.75	8.78
0.5 – 0.25	1.09	2.06	2.68	5.62	12.90	20.10	10.10	9.95	9.90
0.25 – 0.106	1.15	2.02	1.52	2.79	8.79	12.40	31.30	29.40	7.39
0.106– 0.053	1.12	2.05	0.57	1.38	1.70	4.09	8.77	5.91	10.30
0.053– 0.053	1.01	1.97	0.40	0.77	1.37	2.14	10.80	3.99	9.61
Geometric mean diam (mm)			9.06	3.03	1.08	0.68	0.52	0.81	1.81
			Bulk density, g cm^{-3}						
Measured			1.09	1.15	1.27	1.39	1.46	1.50	1.44
Predicted			1.08	1.20	1.26	1.32	1.33	1.33	1.35

of each aggregate fraction was measured using a similar packing procedure except in a smaller container. Bulk density of individual aggregates was determined from weight and volume measurements. The volume of large aggregates was measured from the displacement of water by aggregates after they had been coated with wax (Blake, 1965) or enamel (Gupta, S. C., and W. E. Larson. 1980. Determining density of soil aggregates. Unpublished manuscript). The volume of small aggregates was estimated from the displacement of glycerine by kerosene-saturated aggregates (Gupta, S. C., and W. E. Larson. 1980. Determining density of soil aggregates. Unpublished manuscript).

Water retention characteristic curves of Webster clay loam at several bulk densities were determined by the procedure outlined by Gupta and Larson (1979b).

Experiment 3

Changes in the total porosity of soil as influenced by water content and applied stress were calculated from Eq. [1] and the relationship described below:

$$S_1 = \frac{100w\varrho}{f\,\varrho_w} \qquad [8]$$

$$f = \left(1 - \frac{\varrho}{\varrho_p}\right)100 \qquad [9]$$

where
w = water content by weight, %.
ϱ_w = density of water = 1 g cm⁻³.
f = total porosity, %.
ϱ_p = soil particle density = 2.6 g cm⁻³.

Using the data from compression and pore water pressure changes studies (Larson et al., 1980; Larson and Gupta, 1980) on 55 soils, a relationship was developed between the parameters of Eq. [1] and the particle size analysis. Particle size was determined by the international pipette method (Day, 1965).

Experiment 4

Soil water contents and vehicular loads which are not conducive to detrimental soil compaction at the time of tillage were determined after superimposing the following criteria on the porosity vs. water content diagrams obtained in Exp. 3: (1) air-filled porosity (10%) that limits the gaseous diffusion between the soil and the atmosphere, (2) stress values above which the soil aggregates will shear, and (3) the penetration resistance (σ_p = 20 kg cm⁻²) that limits the root elongation.

Critical stress was estimated using the pore water pressure curves and the procedure described by Larson and Gupta (1980). Penetration resistance was measured by forcing a penetrometer (Diameter = 0.376 cm, α = 30°) through the soil cores (7.6-cm diam × 7.6-cm high) remolded at several water contents and bulk densities. The rate of penetration (3.0

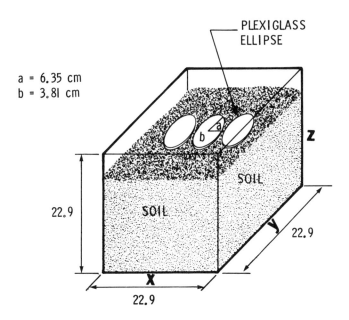

Fig. 2. Schematic of load applications on the soil surface in a box. An elliptical plexiglass plate simulated the shape of the contact area between the tire and the soil.

cm/hour) was controlled by the movement of the crossbar in the Instron Universal Testing Machine (IUTM).[3] Remolded cores were made after compressing the loose wet soil to a desired bulk density on the IUTM. Ground soils passed through a 2-mm sieve and wetted with an atomizer were used for remolding.

Experiment 5

The concentration factor, v, in Frohlich's equations (Eq. [2], [3], and [4]) was evaluated in the laboratory. Moist soil was packed in a clear plexiglass cube with a side length of 22.9 cm. An elliptical shaped (a = 6.35 cm, b = 3.81 cm) plexiglass plate simulated the shape of the contact area between a tire and soil (Soehne, 1953). The continuous path of the tractor tire was simulated by applying the load to the plexiglass plate at three locations on the surface of soil in the box. Figure 2 is a schematic of the load applications. The first load was applied in the center of the soil surface whereas the second and the third loads were applied on either side of the imprint left by the first load. Soil in the box was scanned with a gamma probe in the y and z-directions, both before and after the application of loads. Bulk densities were estimated from the gamma count and the calibration curves. Soil in the box was scanned at the 2.5-cm interval except for later runs when the scan was done in 1-cm intervals near the soil surface.

The concentration factor, v, was estimated by matching the middle part of the iso-bulk density lines from the measured data with that of the predicted bulk densities at several θ values. Predicted densities were obtained by using the predicted stress values (procedure suggested by Soehne (1953) using Eq. [2], [3], and [4]) in the compression model (Eq. [1]) of Larson et al. (1980).

Experiment 6

Soehne's procedure for predicting iso-stress lines along with the compression model (Eq. [1]) was also used to simulate the effect of a tractor traffic (Load = 2,045 kg) on the soil bulk density profile. Data on tire size, contact area, and tire pressure were taken from Vanden Berg et al. (1957). Parameters of Eq. [1] were estimated from particle size analysis and the relationships developed in Exp. 3 (Eq. [10] through [14]). The soil profile was delineated into three zones: (a) areas where stresses are above the critical limit of shearing aggregates, (b) areas where air-filled porosity is limiting for gaseous diffusion, and (c) areas where soil-root friction is critical for root growth.

Briefly, the inputs needed to use the compaction model are (a) the load on the tractor tire, (b) the axis of the ellipse describing the contact area between the tire and the soil surface, (c) the clay type, (d) the desired

---

[3]Mention of trade names is for the convenience of the reader only and does not imply endorsement by the USDA over similar products of other companies not mentioned.

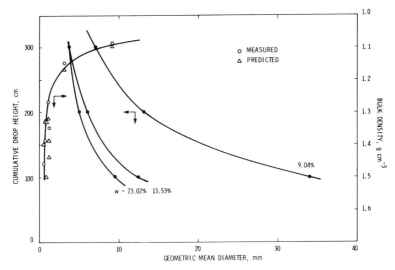

Fig. 3. Relationship between cumulative drop height, packing densities, and the geometric mean diameter of the aggregates from Webster clay loam. The w is water content by weight.

degree of water saturation at $\sigma_a = 1$ kg cm^{-2}, and (e) the relationships (similar to Fig. 12 and 13) for estimating the bulk density values corresponding to critical penetration resistance.

RESULTS AND DISCUSSION

Energy Requirement for Soil Breakup

Figure 3 shows the relationship between the cumulative drop height and the geometric mean diameter of the shattered aggregates at three water contents. Cumulative drop height is an index of the shattering energy whereas geometric mean diameter is an index of aggregate size distribution. Geometric mean diameter of shattered soil aggregates decreased exponentially with an increase in the cumulative drop height. An exponential relationship indicates a small change in the geometric mean diameter of shattered aggregates with an increase in the cumulative drop height. In other words, more energy is needed to further break up unit mass of small aggregates than the large aggregates. Dropping the clods of higher water content resulted in the smaller geometric mean diameter of the shattered aggregates at the same cumulative drop height. This means that more energy is required for breaking a unit mass of dry than of wet soil to achieve the same aggregate size distribution.

A relationship similar to Fig. 3 can be developed between the field measurements of the breakup energy from the tillage implements and the aggregate size distribution. This has been shown by Gill and McCreery

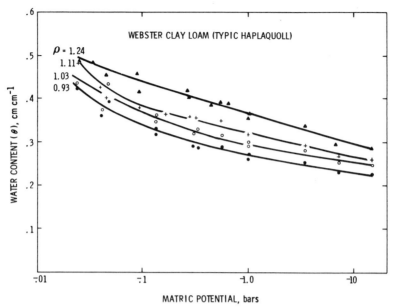

Fig. 4. Water retention characteristics of Webster clay loam at four bulk densities. Water retention is expressed on a volume basis, θ.

(1960) who used the draft forces as an equivalent of breakup energy. Draft forces varied with the size of the cut and of the tillage implement in their experiment.

Bulk Density Prediction of Aggregated Soils

Table 1 shows the input used for predicting the packing bulk density of aggregated Webster clay loam. Also given in Table 1 are the predicted and the measured packing bulk density for seven different aggregate size distributions. Each one of the seven distributions is dominated by a certain group of aggregate fractions. This is reflected in the geometric mean diameter of these distributions.

Packing densities when plotted (Fig. 3) against the geometric mean diameter showed an exponential increase with decrease in the geometric mean diameter. In general, the predicted packing bulk densities are close to the measured values.

Soil Breakup Model

The soil breakup model is based on the assumptions that (a) the optimum soil physical (water storage, heat flow, and gaseous diffusion) or biological (microbial activity, seed germination, and seedling growth) processes are associated with a particular bulk density of the soil; (b) the

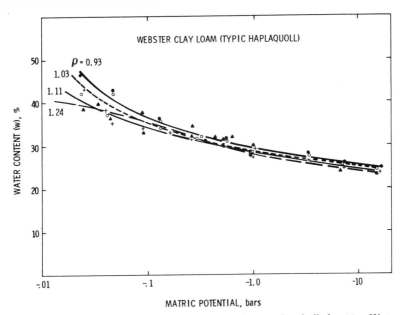

Fig. 5. Water retention characteristics of Webster clay loam at four bulk densities. Water retention is expressed on a weight basis, w.

packing bulk density of medium to fine-textured agricultural soil is related to the aggregate size distribution and the aggregate characteristics; and (c) the energy required to create different aggregate size distribution varies with the water content, the soil, and the tillage implement. Thus, if the optimum bulk density is known, then the procedure suggested in the soil breakup model can be used to select tillage implements suited for a soil at a given water content that would provide an aggregate size distribution optimum for a soil process.

For example, Fig. 4 shows the water retention characteristics of Webster clay loam at four bulk densities. At a given matric potential, the water retention increases with an increase in the bulk density of the soil. If the water content is expressed on a weight basis, the retention curves of Fig. 4 look like Fig. 5. Below 31% water content by weight or a potential of less than −0.33 bar, all the water retention curves are approximately the same. We believe that above approximately −0.33 bar matric potential in Webster clay loam soil, the majority of the water is held in the pore spaces between the aggregates. Similarly, below approximately −0.33 bar matric potential, the majority of the water is held in the intra-aggregate pore spaces, in the inter-aggregate pores that are the same size as the intra-aggregate spaces, and absorbed on the surface of the finer particles (clay, organic matter). The dividing point (w = 31% or −0.33 bar matric potential for Webster clay loam) in the water retention curve would vary with the type of soil depending upon the porosity and swelling and shrinking characteristics of its aggregates. Since the retention characteristics of different size aggregates in a soil are approximately the same

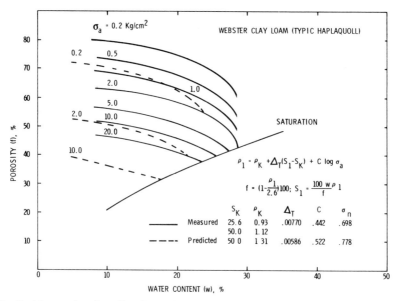

Fig. 6. Measured and predicted porosity vs. water content relationship of a Webster clay loam at various applied stresses.

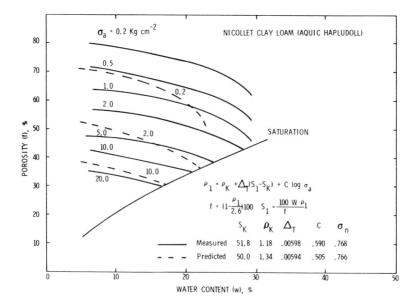

Fig. 7. Measured and predicted porosity vs. water content relationship of Nicollet clay loam at various applied stresses.

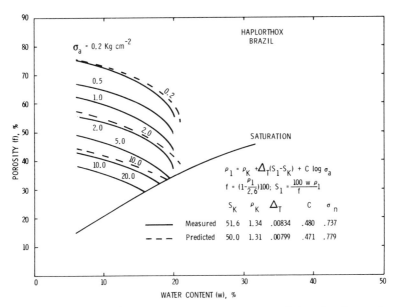

Fig. 8. Measured and predicted porosity vs. water content relationship of a sandy soil (Haplorthox) at various stresses.

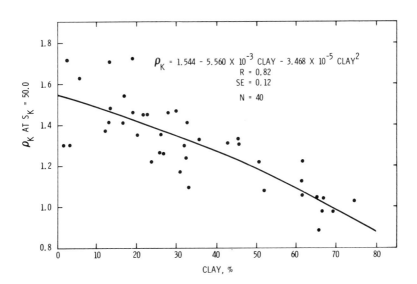

Fig. 9. Relationship between ϱ_K and percent clay.

(Gupta, S. C., and W. E. Larson. 1981. unpublished data) (partially evident from nearly the same water retention characteristics below -0.33 bar matric potential at different bulk densities in Fig. 5), the retention characteristic of a soil could only be changed by changing the pore space distribution between the aggregates. In turn, this could be achieved by creating different aggregate size distributions through management practices like tillage.

Thus, if the objective of a management is to reduce evaporation (liquid soil water) and increase infiltration in a dry year, seed beds with large voids may be more appropriate. Soils with large voids are those with a bulk density less than 1.1 g cm^{-3}. These bulk densities correspond (Fig. 3) to an aggregate size distribution with a geometric mean diameter greater than 8.3 mm. From Fig. 3 the equivalent breakup energy for the above distribution is equal to a cumulative drop height of less than 138 cm at 13.5% water content by weight. It is possible that this energy can be achieved through several tillage implements. Thus, if the efficiency of the tillage tool is known, the implement can be selected to achieve the required aggregate size distribution. An example of this has been given by Gill and McCreery (1960).

In summary, the soil breakup model involves (a) deciding the bulk density of soils that is optimum for a soil process, (b) estimating aggregate size distribution from the packing model that gives the approximate desired bulk density, (c) estimating the breakup energy needed to create the desired aggregate size distribution, and (d) selecting the tillage implement that provides the needed breakup energy at a given water content. The last two steps would involve determining relationships between the field-measured breakup energy and the geometric mean diameter of the resultant aggregate size distribution.

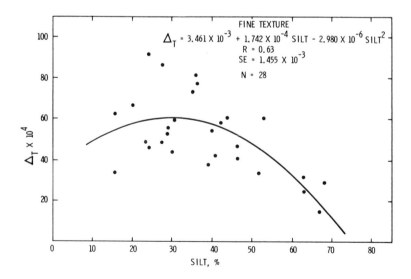

Fig. 10. Relationship between Δ_T and percent silt for fine-textured soils.

Predicting Changes in the Soil Porosity as Influenced by the Water Content and the Applied Mechanical Stress

Figures 6, 7, and 8 show the comparisons between the measured and the predicted porosity as influenced by water content and applied mechanical stress for Webster clay loam, Nicollet clay loam, and a sandy soil (Haplorthox) from Brazil. Both the measured and the predicted porosities were estimated from Eq. [1], [8], and [9]. For measured porosities, the value of variables in Eq. [1] were the measured values obtained from the uniaxial compression curves of the disturbed soil samples (Larson et al., 1980; Larson and Gupta, 1980). In the predicted relationships, the value of the parameters in Eq. [1] were estimated from the particle size analysis and the relationships developed in this study. These relationships as described below were developed from the basic data of 54 samples of Larson and Gupta (1980) and Webster clay loam used in this study.

ESTIMATION OF ϱ_K

Figure 9 shows the relationship between ϱ_K and the percent clay. Since measured ϱ_K corresponded to a different degree of saturation for different soils, all values of ϱ_K used in the regression equation were standardized to a reference degree of saturation (50%) by interpolation between the measured values. The Regression equation is:

$$\varrho_K = 1.544 - 5.560 \times 10^{-3} \, (\% \text{ clay}) - 3.468 \times 10^{-5} \, (\% \text{ clay})^2$$

$$\text{at } S_k = 50\% \qquad\qquad R = 0.82 \qquad\qquad [10]$$

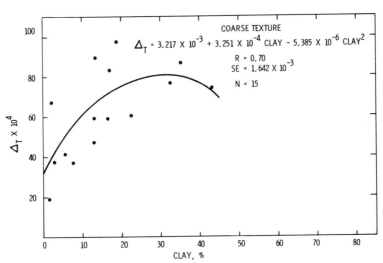

Fig. 11. Relationship between Δ_T and percent clay for coarse-textured soils. Soils included in the coarse-textured group were sand, loamy sand, sandy loam, sandy clay loam, and sandy clay.

ESTIMATION OF Δ_T

Stepwise regression analysis gave the best correlation coefficient of Δ_T with percent silt for the fine-textured soils (Fig. 10) and Δ_T with percent clay for the coarse-textured soils (Fig. 11). Regression equations that described these relationships are:

$$\Delta_T = 3.461 \times 10^{-3} + 1.742 \times 10^{-4}\,(\%\ \text{silt}) - 2.980 \times 10^{-6}\,(\%\ \text{silt})^2$$

$$R = 0.63 \tag{11}$$

for the fine-textured soils and

$$\Delta_T = 3.217 \times 10^{-3} + 3.251 \times 10^{-4}\,(\%\ \text{clay}) - 5.385 \times 10^{-6}\,(\%\ \text{clay})^2$$

$$R = 0.70 \tag{12}$$

for the coarse-textured soils.

Soils included in the coarse-textured group and Fig. 11 were sand, loamy sand, sandy loam, sandy clay loam, and sandy clay. All others were in the fine-textured group.

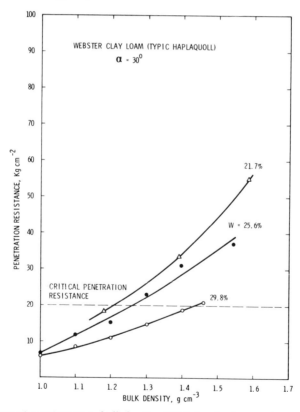

Fig. 12. Penetration resistance vs. bulk density of Webster clay loam at three water contents. The w is water content by weight.

ESTIMATION OF C

Larson et al. (1980) showed that compression index, C, increases up to a clay content of about 33% and then levels off. Regression equations that described the relationships between C and percent clay are:

$$C = 2.033 \times 10^{-1} + 1.423 \times 10^{-2}\,(\%\ \text{clay}) - 1.447 \times 10^{-4}\,(\%\ \text{clay})^2$$

$$R = 0.79 \qquad [13]$$

for the temperate region soils with expanding type clays and

$$C = 1.845 \times 10^{-1} + 1.205 \times 10^{-2}\,(\%\ \text{clay}) - 1.108 \times 10^{-4}\,(\%\ \text{clay})^2$$

$$R = 0.89 \qquad [14]$$

for the tropical and semi-tropical region soils with nonexpanding type clays. Equations [13] and [14] contain the data of additional soils and are thus slightly different than those of Larson et al. (1980). Maximum C

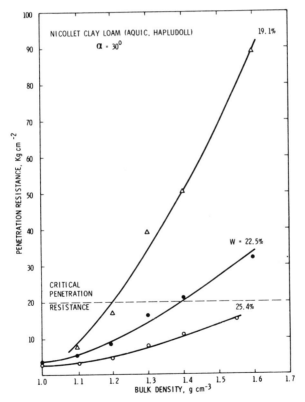

Fig. 13. Penetration resistance vs. bulk density of Nicollet clay loam at three water contents. The w is water content by weight.

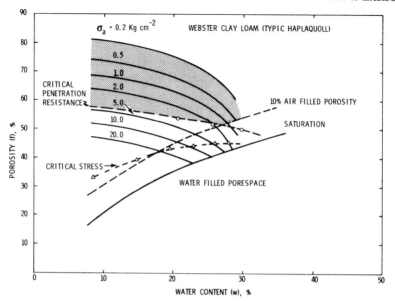

Fig. 14. Porosity vs. water content diagram of Webster clay loam with superimposed criteria of (1) 10% air-filled porosity, (2) stress critical for shearing aggregates, and (3) critical penetration resistance.

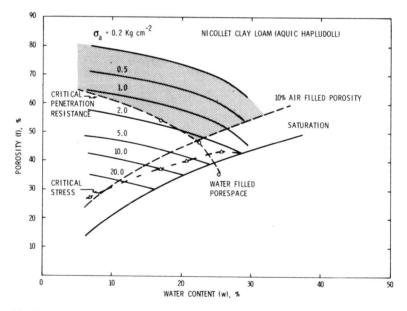

Fig. 15. Porosity vs. water content diagram of Nicollet clay loam with superimposed criteria of (1) 10% air-filled porosity, (2) stress critical for shearing soil aggregates, and (3) critical penetration resistance.

value was 0.59 for soils with predominantly expanding type clays and 0.56 for soils containing predominantly nonexpanding type clays.

Equations [10] through [14] along with Eq. [1] provide a means to approximate compression behavior of other soils from their particle size analysis and the clay type. Predicted compression curves can then be used to estimate the changes in the porosity as influenced by the water content and the applied mechanical stress (Eq. [8] and [9]). Figures 6, 7, and 8 show the differences between the predicted and measured porosities at several water contents and mechanical stresses. Measured and predicted porosities greatly differ for Webster clay loam (Fig. 6) followed by Nicollet clay loam (Fig. 7) and the sandy soil from Brazil (Fig. 8). Comparison of measured and predicted values of the parameters of Eq. [1] (inset tables in Fig. 6, 7, and 8) showed differences in the measured and predicted values of ϱ_K as the main reason for the large differences between the measured and predicted porosities at a given water content and applied mechanical stress. The other parameter that contributed to this difference was the estimate of Δ_T. Although compression studies were done (Larson and Gupta, 1980) on 55 samples, the values of the Δ_T and ϱ_K were not available on samples with less than three compression curves. This led to a poor correlation coefficient of these parameters with the particle size analysis (Fig. 9, 10, and 11). In general, all the relationships in Eq. [10] through [14] need further improvement which can be achieved by running additional compression studies on other soils.

Prediction of Water Contents and Vehicular Stress not Conducive to Detrimental Levels of Soil Compaction

Porosity vs. water content diagrams similar to Fig. 6, 7, and 8 can be used to estimate a range of water content above which a soil is susceptible to detrimental compaction by stresses from tillage equipment. Three criteria used for soil compaction are (1) air-filled porosities critical for gaseous diffusion, (2) stresses critical for shearing soil aggregates, and (3) soil resistance critical for root growth.

We assumed 10% air-filled porosity as a critical value below which gaseous exchange with the atmosphere may restrict biological activities such as root respiration. This, however, depends upon the type of plant and the level of microbial activity in the soil.

Critical stresses were estimated from the measured pore water potential vs. applied stress curves obtained during the compression tests (Larson and Gupta, 1980). Critical stress could also be estimated from the particle size analysis using Eq. [5] and the relationship described below

$$\sigma_n = 5.517 \times 10^{-1} + 10.2 \times 10^{-3} \, (\% \text{ clay}) - 10.6 \times 10^{-5} \, (\% \text{ clay})^2$$

$$R = 0.81 \tag{15}$$

Details of this procedure have been described by Larson and Gupta (1980).

Penetration resistance of 20 kg cm^{-2} was assumed critical for root penetration. This corresponds to a normal point resistance of approximately 15 kg cm^{-2} for fine-textured soils (Voorhees et al., 1975). However, the critical level of penetration resistance will vary with the type of root system, penetrometer, and soil. Porosities corresponding to critical penetration resistance at three water contents were estimated from the laboratory-obtained curves (Fig. 12 and 13) on remolded soil cores.

These three criteria superimposed on the measured porosity vs. water content curves of Fig. 6 and 7 are shown in Fig. 14 and 15. Both these figures show that the critical stress curve crosses over the 10% air-filled porosity curve at approximately $\sigma_a = 20$ kg cm^{-2}. This indicates that if at least 10% air-filled porosity is maintained after compaction, applied stress ($\sigma_a = <20$ kg cm^{-2}) from the vehicular traffic is not critical for shearing soil aggregates. Generally, the stress due to agricultural traffic on the farm are less than 5.0 kg cm^{-2}. Trends for other soils tested in this study were similar to Fig. 14 and 15. Thus, 10% air-filled porosity is one limit that can be used to judge acceptable range of water content and

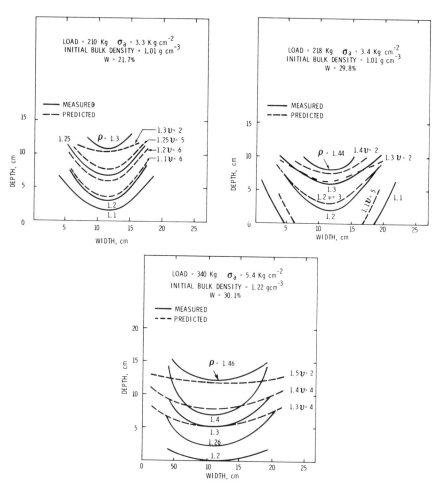

Fig. 16. Measured and predicted iso-bulk density lines in Webster clay loam under simulated wheel tracks.

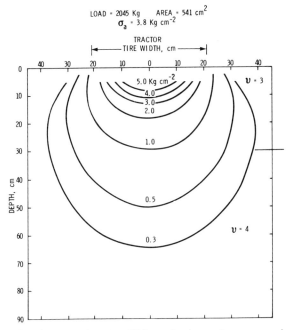

Fig. 17. Predicted iso-stress lines in a Webster clay loam after a passage by a tractor load.

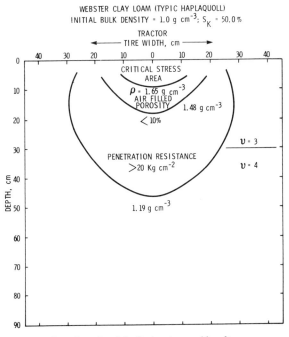

Fig. 18. Delineation of predicted soil bulk density profile after a passage of a tractor into zones based on (1) stress critical for shearing aggregates, (2) limiting air-filled porosity, and (3) critical penetration resistance.

stress which do not create detrimental level of soil compaction. Another limit that dictates the range of water content and applied stress by vehicular traffic at the time of tillage is the penetration resistance (Fig. 14 and 15). For example, Fig. 15 shows that the penetration resistance of Nicollet clay loam at 20% water content by weight would reach a critical limit (20 kg cm^{-2}) for root growth if the applied compressive stress is greater than 2.0 kg cm^{-2}. On the other hand, the corresponding stress for Webster clay loam is 3.0 kg cm^{-2} (Fig. 14). The lower value for Nicollet clay loam indicates its greater susceptibility to compaction as compared to Webster clay loam at the same water content.

The concept of critical penetration resistance is useful as long as the field water contents are less than the water content at the time of tillage. For example, if at the time of tillage the Webster clay loam (21.7% water content) is compacted to a bulk density of 1.21 g cm^{-3}, the critical penetration resistance is reached (Fig. 12). Thus, during the growing season, if the field water content is less than 21.7%, the penetration resistance will be higher than the critical level for root growth. On the other hand, if the water content is higher than 21.7% during the growth season, the penetration resistance will be lower than the critical limit (Fig. 12).

The shaded area in Fig. 14 and 15 gives the combination of water contents and traffic stresses that prohibit excessive soil compaction at the time of tillage.

Calibration of Frohlich's Equations

Measured and predicted bulk densities with depth for two water contents and two applied stresses under simulated wheel tracks are shown in Fig. 16. For all conditions studied in this laboratory experiment, v values that gave the best prediction of iso-bulk density line were small near the surface and increased with soil depth. These findings are similar to the observation of Blackwell and Soane (1980) but are in contrast to Soehne's recommended values $v = 4$ for hard or dry soil, $v = 5$ for firm or average moisture condition, and $v = 6$ for soft and wet soil. Differences between Soehne's recommendations and our findings may be because our compression model (Eq. [1]) was developed on thin samples with multiple passes of an applied load, whereas our simulated wheel tracks were on a large soil sample in a box with a single application of an applied load.

Soil Compaction Model

Figure 17 shows the predicted distribution of iso-stress lines in the soil after passage of a tractor with an applied stress of 3.8 kg cm^{-2} at the soil surface (Exp. 6). In Fig. 17, iso-stress lines above and below the 30-cm soil depth represent $v = 3$ and $v = 4$, respectively, in the Frohlich equations (Eq. [2], [3], and [4]). Selection of 30 cm as the boundary was arbitrary. More experiments will be needed to estimate the depth where the v value changes.

A bulk density profile (Fig. 18) of a Webster clay loam soil was calculated from Eq. [1] and the iso-stress lines in Fig. 17. All variables of Eq. [1] were estimated from particle size analysis and Eq. [10] through [14].

Desired degree of saturation was assumed to be 50% at the time of tillage which translates to a water content of 19% by weight. Bulk density before tillage traffic was assumed to be 1.0 g cm^{-3}. Using the porosity vs. water content diagram (Fig. 6), the bulk density profile (Fig. 18) was delineated into three zones of limiting conditions: (a) in the 0 to 9-cm depth, the applied mechanical stress was higher than the critical value for shearing soil aggregates; (b) in the 0 to 18-cm depth, the air-filled porosity was below 10% and thus critical for gaseous exchange; and (c) in the 0 to 45-cm depth, the penetration resistance was higher than the critical limit for root growth. The depth of critical penetration resistance and 10% air-filled porosity will, however, change as the water content changes during the growing season. Estimated boundaries of the three zones in Fig. 18 reflect the error of estimating ϱ_K from particle size analysis. To avoid this error an alternative is to measure ϱ_K for a given soil at a given water content.

The computation for Fig. 18 illustrates the procedure of using the particle size analysis to predict the bulk density profile of a soil after a passage of tillage traffic. It further shows how these profiles can be delineated into three zones based on conditions that limit plant growth and degrade soil structure. The compaction model also contributes to our understanding of soil deformation and suggests ways to alleviate the compaction problem. One possibility is to use controlled traffic and avoid destruction of soil structure year after year and all over the field.

AREAS OF FUTURE RESEARCH

Soil Breakup Model

1. Field measurements of soil breakup energy from the tillage implements and the GMD of the shattered aggregates.
2. Generalizing these relationships by textural class.
3. Further testing of the packing model on aggregated soils.
4. Generalizing the measured aggregate characteristics by textural class.
5. Defining bulk density values that optimize various soil physical (water storage, heat flow, gaseous diffusion) or biological (seed germination, seedling growth, microbial activity) processes.
6. Field testing of the soil breakup model predictions.

Soil Compaction Model

1. Compression tests on additional soils.
2. Refinement of the regression equations (Eq. [10] through [14]) that describe the components of the compression model (Eq. [1]).
3. Estimating v-values from field and laboratory experiments.
4. Delineating profile depths where v changes.
5. Developing and generalizing by textural class, the relationships that describe the resistance of soils to the penetration of steel probe.

6. Defining critical levels of the penetration resistance and air-filled porosities based on root growth and gaseous diffusion, respectively.

7. Field testing the model predictions on the distribution of stress and bulk density in a soil profile after passage of vehicular traffic.

DEFINITION OF SYMBOLS

C = compressive index (slope of ϱ vs. log σ_a curve)

GMD = geometric mean diameter, mm

MWD = mean weight diameter, cm

N = number of observations

P = load at a point, kg

R = correlation coefficient

S = degree of saturation, %

SE = standard deviation of the regression

S_K = degree of saturation at $\sigma_a = 1$ kg cm^{-2}, %

S_1 = desired degree of saturation at σ_K, %

U_m = minimum pore water potential, bars

W_i = weight of aggregates in a sieve class i, gm

a = major semiaxis of the ellipse, cm

b = minor semiaxis of the ellipse, cm

f = porosity, %

r = radial distance in the soil profile from point 0, cm

w = water content by weight, %

Υ = shearing stress belonging to σ_z and σ_h, kg cm^{-2}

σ_K = reference applied stress = 1 kg cm^{-2}

σ_N = normal point resistance, kg cm^{-2}

σ_P = total point resistance, kg cm^{-2}

σ_a = applied compressive stress, kg cm^{-2}

σ_c = critical stress or σ_a corresponding to U_m

σ_h = horizontal normal stress in a radial direction, kg cm^{-2}

σ_n = normalized stress, log σ_a/log σ_s, at U_m

σ_s = applied stress at saturation, kg cm^{-2}

σ_r = stress in a volume element of the soil, kg cm^{-2}

σ_z = vertical normal stress, kg cm^{-2}

ϕ' = coefficient of soil metal friction

α = included semiangle of the conical probe, degree

β = angle bisected by a vertical line from point 0 with a line to the center of gravity of the volume element under question

υ = concentration factor

ϱ = bulk density, g cm^{-3}

ϱ_K = bulk density at 1 kg cm^{-2}, g cm^{-3}

ϱ_a = bulk density of individual aggregate, g cm^{-3}

ϱ_b = bulk density of aggregate fraction, g cm^{-3}

π_p = soil particle density = 2.6 g cm^{-3}

π_w = density of water = 1.0 g cm^{-3}

Δ_T = slope of ϱ_K vs. S curve

LITERATURE CITED

1. Allmaras, R. R., R. E. Burwell, W. B. Voorhees, and W. E. Larson. 1965. Aggregate size distribution in the row zone of tillage experiments. Soil Sci. Soc. Am. Proc. 29:645–650.

2. Barnes, K. K., W. M. Carleton, H. M. Taylor, R. I. Throckmorton, and G: E. Vanden Berg. 1971. Compaction of agricultural soils. Am. Soc. Agric. Eng., St. Joseph, Mich. 471 p.

3. Bateman, H. P., M. P. Naik, and R. P. Yoerger. 1965. Energy required to pulverize soil at different degrees of compaction. J. Agric. Eng. Res. 10:132–141.

4. Blackwell, P. S., and B. D. Soane. 1981. A method of predicting bulk density changes in field soils resulting from compaction by agricultural traffic. J. Soil Sci. 32:51–65.

5. Blake, G. R. 1965. Bulk density. In C. A. Black et al. (ed.) Methods of soil analysis. Part 1. Agronomy 9:374–390. Am. Soc. of Agron., Madison, Wis.

6. Chancellor, W. J. 1976. Compaction of soil by agricultural equipment. Div. of Agric. Sci. Bull. 1881. Univ. of California, Richmond, Calif.

7. Day, P. R. 1965. Particle fractionation and particle size analysis. In C. A. Black et al. (ed.) Methods of soil analysis. Part 1. Agronomy 9:545–567. Am. Soc. of Agron., Madison, Wis.

8. Farrell, D. A., E. L. Greacen, and W. E. Larson. 1967. The effect of water content on axial strain in a loam soil under tension and compression. Soil Sci. Soc. Am. Proc. 31:445–450.

9. Frohlich, O. K. 1934. Druckverteilung in Baugrunde (Pressure distribution in the soil). Wien.

10. Gill, W. R., and W. F. McCreery. 1960. Relation of size of cut to tillage tool efficiency. Agric. Eng. 41:372–374.

11. Grable, A. R. 1971. Effects of compaction on content and transmission of air in soils. p. 154–164. In Compaction of agricultural soils. Am. Soc. Agric. Eng., St. Joseph, Mich.

12. Gupta, S. C., and W. E. Larson. 1979a. A model for predicting packing density of soils using particle size distribution. Soil Sci. Soc. Am. J. 43:758–764.

13. ―――, and ―――. 1979b. Estimating soil water retention characteristics from particle size distribution, organic matter percent, and bulk density. Water Resour. Res. 15:1633–1635.

14. Harris, W. L. 1971. Methods of measuring soil compaction. p. 9–44. In Compaction of agricultural soils. Am. Soc. Agric. Eng., St. Joseph, Mich.

15. Larson, W. E., and S. C. Gupta. 1980. Estimating critical stresses in unsaturated soils from changes in pore water pressure during confined compression. Soil Sci. Soc. Am. J. 44:1127–1132.

16. ―――, ―――, and R. A. Useche. 1980. Compression of agricultural soils from eight soil orders. Soil Sci. Soc. Am. J. 44:450–457.

17. ―――, and J. B. Swan. 1970. Tillage of wet and dry soils. Crops Soils 22:8–11.

18. Marshall, T. J., and J. P. Quirk. 1950. Stability of structural aggregates of dry soil. Aust. J. Agric. Res. 1:260–275.

19. Soehne, W. H. 1953. Pressure distribution in the soil and soil deformation under tractor tires. Grundlagen der Landtechnik 5:49–63.

20. Taylor, H. M., and R. R. Bruce. 1968. Effect of soil strength on root growth and crop yield in the Southern United States. 9th Int. Congr. Soil Sci. Soc. Trans. 1:803–811.

21. ―――, and H. R. Gardner. 1963. Penetration of cotton seedling taproots as influenced by bulk density, moisture content, and strength of soil. Soil Sci. 96:153–156.

22. Taylor, J. H., E. C. Burt, and A. C. Bailey. 1978. Traction and compaction of big tractors. Paper No. 78-1029, Am. Soc. Agric. Eng., St. Joseph, Mich.

23. Vanden Berg, G. E., A. W. Cooper, A. E. Erickson, and W. M. Carleton. 1957. Soil pressure distribution under tractor and implement traffic. Agric. Eng. 38:854–855, 859.

24. Vomocil, J. A., and W. J. Flocker. 1961. Effect of soil compaction on storage and movement of soil, air, and water. Trans. Am. Soc. Agric. Eng. 4:242–246.

25. Voorhees, W. B., D. A. Farrell, and W. E. Larson. 1975. Soil strength and aeration effects on root elongation. Soil Sci. Soc. Am. Proc. 39:948–953.

26. Wesseling, J., and W. R. van Wijk. 1957. Soil physical conditions in relation to drain depth. In J. N. Luthin (ed.) Drainage of agricultural lands. Agronomy 7:461–504. Am. Soc. of Agron., Madison, Wis.

Chapter 11

Predicting Tillage Effects on Cotton Growth and Yield[1]

F. D. WHISLER, J. R. LAMBERT, AND J. A. LANDIVAR[2]

ABSTRACT

Tillage affects several soil properties and these may in turn influence root growth, shoot growth, and ultimately crop yield. Therefore, a crop growth model intended to simulate tillage effects on yield must operate at the physical/physiological process level and must contain appropriate linkages and feedback mechanisms. GOSSYM is such a simulation model for cotton. It takes into account root growth, soil water movement, temperature and mechanical impedance, and after recent modification, accounts for soil oxygen status. This paper describes how the model has been adapted to account for the effects of tillage and wheel traffic on the soil properties affecting water movement, oxygen, and mechanical impedance. Since this model starts at plant emergence, the main tillage practice considered is cultivation. Not only are the effects of cultivation noticeable on soil properties, but they also directly affect the plant by root pruning. The overall effects are demonstrated through changes in the pattern of soil water, oxygen content, root distribution, and crop yield.

[1] Contribution from the Dep. of Agronomy, Mississippi Agric. and For. Exp. Stn., P.O. Box 5248, Mississippi State, MS 39762; Paper No. 4653; and Technical Contribution 1867 of the South Carolina Agric. Exp. Stn., Clemson Univ.

[2] Professor of soil physics, MAFES; Professor of agric. eng., Clemson Univ.; and research associate, MAFES, respectively.

INTRODUCTION

Previous papers of this Symposium have presented the theory and some models on the ways that tillage affects soil physical properties. A plant growing in such a system integrates all of these soil property changes, direct effects of tillage such as root pruning, and weather. A simulation model of such a system, therefore, must have sufficient linkage and feedback relationships to describe the important changes brought about by tillage imposed on the "normal" plant growth before it can simulate the observed responses. The GOSSYM cotton-growth simulation model has been modified to incorporate the effects of tillage on soil properties and root growth. The model has sufficient feedback mechanisms built into it that the plant responses appear reasonable when compared to field or controlled environment observations.

Theory

The GOSSYM model has been presented to the American Society of Agronomy previously (Baker et al., 1979a). Detailed sections of the model and tests of the model have also been presented at the ASA meetings and elsewhere (Lambert et al., 1975; Baker et al., 1976, Lambert et al., 1976, Whisler et al., 1977; Baker, 1979; Baker et al., 1979b, 1979c; Fye et al., 1979; Landivar et al., 1979; Whisler et al., 1979a, 1979b). Therefore, only the detail of the model necessary for describing the tillage and associated processes will be described herein. A flow diagram of the GOSSYM model is given in Fig. 1.

SOIL HYDRAULIC PROPERTIES

The water content-matric potential (pressure head) relationships and water content-diffusivity relationships, were taken from Brooks and Corey (1964) and Gardner and Mayhugh (1958), respectively. They are:

$$(\theta_i - \theta_r)/(\theta_s - \theta_r) = (\Psi_B/\Psi_i)^{(\eta-2)/3} \qquad \Psi_i < \Psi_B \qquad [1a]$$

$$\theta_i = \theta_s \qquad \Psi_i \geq \Psi_B \qquad [1b]$$

and

$$D(\theta_i) = D_o \exp \beta (\theta_i - \theta_o) \qquad [2]$$

where θ_i is the volumetric water content at the matric potential, Ψ_i; θ_s is the saturated (including entrapped air for field values) water content; θ_r is the residual water content; Ψ_B is the air entry value for desorption relationships or the water entry value for absorption relationships; η is a soil characteristic parameter; D_o is the soil water diffusivity at the water content θ_o and β is another soil characteristic parameter. Values of these constants which were used in the simulation are given in Table 1. The methods for estimating the soil characteristic parameters are given in Whisler (1976).

The GOSSYM model assumes that the plant has emerged and starts the simulation at that time. In cotton production, therefore, the main tillage practice to be simulated is cultivation. For much of the cotton grown under natural rainfall conditions it appears from penetrometer readings that the soil settles back from the primary tillage loosening to near its pretilled conditions by the time of emergence, except for traffic pans which are shattered by primary tillage.[3] Therefore, the effects of primary tillage and soil deep disturbance, such as occur after subsoiling, chisel plowing, etc. were ignored. The effects of wheel traffic, however, were taken into account.

In the model the values for each of the soil constants can be changed independently of one another. The following logic was used in the simulation of tillage and wheel traffic effects on soil-water movement but other logic could be used or tested.

1. Cultivation is only 5.05 cm (2 inches) deep.
2. Cultivation does not disturb the soil in a 15 cm (6 inches) zone on either side of the plant.
3. Cultivation changes the soil by increasing the saturated hydraulic conductivity, and thus diffusivity, and the saturated water content (or total porosity). It does not change the residual water content. The other constants in Eq. [1] and [2] are changed accordingly. Table 1 lists the changes in these constants.
4. Subsequent rainfall settles the soil, depending on the rainfall amount, to its precultivated conditions. If the daily rainfall is less than 2.54 mm (0.1 inch), there is no change made to the soil constants.

   For daily rainfall amounts between 2.54 and 25.4 mm (0.1 to 1.0 inch) the soil will settle back to 67% of its original constants' values. A second rainfall of the same amount would settle the soil to 90% of its original constants' values and any further rainfall greater than 2.54 mm would settle it ot exactly its original constants' values. If the first rainfall after a cultivation was greater than 25.4 mm then the soil would be assumed to settle to 90% of its original constants' values, etc.
5. Within the zone of the wheel traffic, the soil properties are uniform, i.e., no horizons.
6. Within the zone of wheel traffic the saturated hydraulic conductivity and saturated water content (total porosity) decrease.
7. The dimensions of the zone of wheel traffic are rectangular.
8. By the time of emergence (start of the simulation), there has been sufficient traffic to establish an equilibrated, zone of traffic, i.e., no changes with time except for possible cultivations.[3]

Assumptions 1 to 4 were incorporated into the cultivation (CULVAT) subroutine, Fig. 1, while items 5 to 8 were established when reading in the initial data.

---

[3]Personal communications. J. R. Williford, Agricultural Engineer, USDA-SEA, Stoneville, MS; L. A. Smith, Agricultural Engineer, USDA-SEA, Auburn, AL and Miss. State, MS.

SOIL MECHANICAL IMPEDANCE

The effects of soil mechanical impedance are calculated in the root impedance (RIMPED) and root growth (RUTGRO) subroutines (Whisler et al., 1977). A relationship between soil penetration resistance and root growth was published by Taylor and Gardner (1963). It is:

$$RG = 104.6 - 3.53\,PR \qquad [3]$$

where RG is the percent root growth compared to nonimpeded growth and PR is penetration resistance in dynes cm⁻². The relationship was found to hold for cotton over a wide range of soil types. The relationship between penetration resistance, soil bulk density, and soil water content

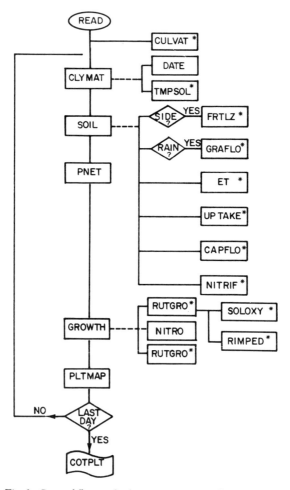

Fig. 1. General flow and subroutine structure of GOSSYM.

Table 1. Soil constants used in the various simulations for a Leeper silty clay loam.

Horizon depth	Variable							
	D_o	θ_o	β	θ_s	θ_r	Ψ_B	η	B.D.
cm	cm²/d	cc/cc		cc/cc	cc/cc	cm		g/cc
0– 16	0.082	0.129	115.2	0.253	0.120	– 60	3.97	1.48
17– 23	0.741	0.178	98.27	0.281	0.170	– 60	3.25	1.85
24– 64	15.44	0.162	42.49	0.331	0.160	– 60	2.45	1.46
65–101	12.45	0.274	60.92	0.390	0.270	– 60	2.42	1.62
Cultivated 0– 5	0.158	0.129	24.60	0.506	0.120	– 60	3.02	1.00
Wheel traffic 0– 31	1.45×10^{-7}	0.129	117.9	0.191	0.120	– 60	2.49	1.85

published by Campbell et al. (1974) was used in these simulations as a table look-up procedure. If this relationship is known for other soils then it could be used. These calculations required the input of the bulk density for each soil horizon to be used in the simulation, Table 1. In order to simulate the effects of tillage upon mechanical impedance the following assumptions were made:
1. Cultivation reduced the bulk density of the surface soil. The bulk density was adjusted for rainfall in the same manner that the hydraulic soil constants were adjusted.
2. The wheel traffic increased the bulk density of the zone of traffic. It was only change in soil surface by cultivation.

Since the soil water content is calculated in the capillary flow (CAPFLO) subroutine, assuming no changes in bulk density with time except as noted above, the penetration resistance can be calculated in the RUTGRO subroutine.

SOIL OXYGEN CONTENT

The effects of soil oxygen content are calculated in the SOLOXY[4] and RUTGRO subroutines. The concentration of oxygen, C, due to one dimensional diffusion into a soil profile may be calculated according to Melhuish et al. (1974) by:

$$C(z,t) = C_b + \alpha z(2L - z)/2D' + (C_i - C_b) 4/\pi \sum_{N=1}^{\infty}$$

$$[1/(2N - 1)(\sin\beta z) \exp(-\beta^2 D't] - (16\alpha L^2/\pi^3 D')$$

$$\sum_{N=1}^{\infty} [1/(2N - 1)^3 (\sin\beta z) \exp(-\beta^2 D' t)] \qquad [4]$$

with the initial condition

$$C(z,o) = C_i(z) \qquad [5]$$

and the boundary conditions

$$C(o,t) = C_b \qquad\qquad [6a]$$

$$(dc/dt)_{L,t} = 0 \qquad\qquad [6b]$$

where C_b is the concentration of oxygen at the upper boundary in g cm^{-3}; α is the soil respiration rate in g cm^{-3}t^{-1}; L is the depth to a gas imperme-able layer in cm; D$'$ is the apparent diffusivity of the gas in cm^2 sec^{-1}, C_i is the initial gas concentration in g cm^{-3} at position z in cm; $\beta = (2N-1)/2L$; N $= 1,2,3\ldots\infty$; and t is time in sec.

A statistical fit of the data of Melhuish et al. (1974) gives

$$D' = -0.0191 + 0.1291\,S \qquad\qquad [7]$$

where S is the air filled porosity. If the total porosity, TP, of a soil horizon is specified then

$$S = TP - \theta_i \qquad\qquad [8]$$

In order to estimate α in [4] the data relating root-soil respiration rate, SOR, to root density, RTD, given in a figure by Mehuish et al. (1974) was fitted statistically as follows:

$$SOR = (0.0305 + 0.0994 \cdot RTD)\,10^{-9} \qquad Z = 30\,cm \qquad [9a]$$

$$SOR = (0.1581 + 0.2014 \cdot RTD)\,10^{-9}] \qquad Z = 90\,cm \qquad [9b]$$

$$SOR = (0.1276 + 0.1809 \cdot RTD)\,10^{-9} \qquad Z = 150\,cm \qquad [9c]$$

In the execution of the program [9a] was used for z $\leq$ 30 cm and then linear interpolation was used between Eq. [9a] to [9c] for other values of z. Note that when RTD $= 0$, SOR is the respiration rate of the soil with-out roots. RTD is calculated in the simulation and thus is used in Eq. [9a] to [9c]. The appropriate value of SOR is then used for α in Eq. [4].

The effect of oxygen concentration on root elongation rate, ELR, is given by Eavis et al. (1971). The data was fitted statistically to give:

$$ELR = 0.0293 - 0.0190\,(c_{z,t}) + 4.8177\,(c_{z,t})^2 \qquad\qquad [10]$$

This equation was used for $c_{z,t} < 0.10$ atm of oxygen (converted from g cm^{-3}). For higher values of $c_{z,t}$ the elongation rate was assumed not to be reduced due to a lack of oxygen, in agreement with these authors. Thus, as cultivation and wheel traffic change the total porosity and hydraulic properties of the soil, they will affect the simulation of the oxygen status of the soil and ultimately the root growth.

SOIL TEMPERATURE

Soil temperature, as it is now calculated in GOSSYM in the TMPSOL subroutine, would not be affected by cultivation or wheel traffic. It is a multiple regression relationship based upon air temperature as determined by McWhorter and Brooks (1965). The regression equation was based upon a full year's data taken at 5, 10, 20, and 40 cm depths as well as air temperatures. Since this is a regression relationship developed for an undisturbed soil, it does not take into account changes in surface soil roughness and conductivity due to cultivation.

METHODS AND MATERIALS

The weather and plant data used in the simulations were from the experiments of Bruce and Romkens (1965). The soil was a Leeper silty clay loam (fine, montmorillonitic, nonacid, thermic Vertic Haplaquepts). Cotton was grown under rainout shelters and well watered until flowering. The high water treatment was irrigated when the matrix potential at 15 cm reached -0.3 bars and received 375 mm of water during the growing season. The low water treatment was irrigated when the matrix potential at 45 cm reached -2.3 bars and received 185 mm of water during the growing season. The weather data are read from a data file by the read (READ) subroutine and converted in the climate (CLYMAT) subroutine. The input data needed are: solar radiation, maximum and minimum daily temperature, rainfall, and the Julian day number. Other data variables not directly used in the cultivation, wheel traffic or oxygen calculations, are listed and their numerical values given:

POPPLT = 20498., LATUDE = 34., EMERGE = 129., SEASON = 170., ROWSP = 91.0, NPLT = 75., NSIDE = 75., APDAY = 28., VARIETY = 0., LTYPE = 0, RUN IDENTIFICATION, TREAT = 1, NOITR = 5, NFRQ = 10, print control variables (self-explanatory at execution), XTR1 = 9.99, XTR2 = -0.55, XTR3 = -5.50, XTR4 = 1.80, A1 = 0.8, A21 = 0.2, A22 = -2.0, A31 = 4.0, A32 = 40.0, CZD = 0.78, CZN = 2.10, CSQ = 2.30, CBL = 0.15, CL = 2.90, CM = 1.20, CPF = 1.20, CVB = 28.0, DPSMX = 0.9, DPBMX = 0.9.

The total porosity, TP, is calculated from θ_s by

$$TP = \theta_s/TPF \qquad [11]$$

The TPF factor can be varied. Baver et al. (1972) indicate that values of TPF between 0.45 to 0.89 are possible for well aerated to poorly aerated soils. For tests of sensitivity with the present data[4], it was found that values of 0.5 and 0.65 represented well aerated and poorly aerated conditions, respectively. TPF is read in after A32 listed above.

The soil properties given in Table 1 are read in along with the depth of the soil horizon, starting with the Ap horizon. First the undisturbed soil properties are read, then those of the cultivated layer and finally those of the zone of wheel traffic. The next data to be read are the table of soil mechanical resistance, water content, and bulk density values. The final input data are the dates (Julian) of the cultivations.

---

[4]Talib, J. B. 1980. The oxygen consumption by soybean roots and soil and its effect on soybean root growth. Unpublished Ph.D. Dissertation. Dep. of Agronomy, Mississippi State University.

The computer output is variable, depending upon what the operator has specified. All of the variables listed above from POPPLT to DPBMX and TPF are listed. The cultivation dates are listed. The number of soil horizons and their values of depth, D_o, θ_o, and β are listed as are the same variables for the cultivated and wheel trafficked areas. The Brooks and Corey constants are not listed. The table of mechanical resistance, water content, and bulk density may be listed if specified. Several plant parameters are named and listed daily (one line per simulated day). Other plant parameters and soil and plant maps are printed 1 day before and at the day of cultivation and at regular intervals, depending upon the value given for NFRQ e.g., 10 will give listings every 10 days. The soil and plant maps are selected by specification of the print control variables at execution time.

The climatic and soils sections of GOSSYM, which are called

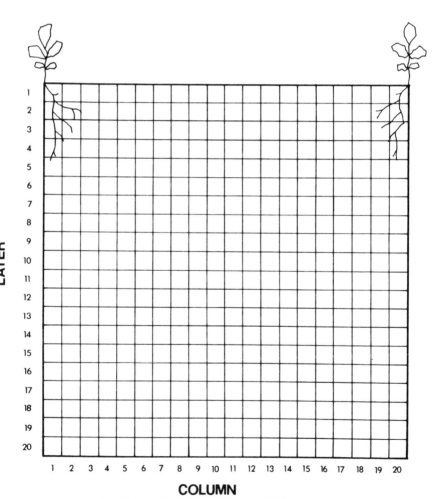

Fig. 2. Spatial configuration of RHIZOS.

RHIZOS (Lambert et al., 1975), work on a two dimensional soils matrix or grid as is shown in Fig. 2. The daily rainfall (or irrigation) enters the soil and is distributed as a one dimension water balance or gravity flow in GRAFLO, i.e., each horizontal layer is wetted until $\theta_i = \theta_s$, if there is sufficient water, before the next lower layer is wet. Water is taken up from each cell in which there are plant roots as calculated in the subroutine UPTAKE and lost from the system as either transpiration or evaporation from the top layer of unshaded soil cells as calculated in the subroutine ET.

The soil water is moved to the root zone and elsewhere in the soil by Darcy flow as calculated in CAPFLO. The root growth from cell to cell is calculated in RUTGRO. Since the water content, mechanical impedance, bulk density, and oxygen content are known for each cell then root growth is reduced from potential root growth due to these factors.

The SOLOXY subroutine first solves for $C(z,t)$ in Eq. [4], using ambient air concentrations of O_2 for C_b, column by column from top to bottom. Advantage is taken of the fact that many of the soil processes, i.e., water flow, root growth, oxygen flow, and heat flow, are symmetrical about the vertical midplane, between the plants of Fig. 2. After the O_2 concentration is calculated in the vertical dimension, then it is calculated in the horizontal direction using the O_2 values from the first calculation in column 1 as C_b for each layer and the midplane as a plane of symmetry or impermeable boundary. This is similar to the alternating direction technique used in numerical analysis.

Simulations were made with or without a 26 cm wide × 31 cm deep zone of wheel traffic midway between the plants, with or without limited aeration problems, i.e., TPF = 0.65 or 0.5, and with or without cultivations at 3, 14, 24, 35, and 50 days after emergence. A three cultivation simulation used cultivations at 3, 24, and 50 days and a four cultivation simulation used all except the 35th or 50th day.

RESULTS AND DISCUSSION

The simulation predictions of cotton lint yield for different cultivation schedules, with and without wheel traffic and with and without limited aeration problems are shown in Table 2. The actual lint yield was 1,728 kg/ha[5]. This would have compared to the no cultivations and no wheel traffic simulations, but there were no measurements of whether or not aeration might have been limited. In general, from Table 2 it is observed that wheel traffic reduced yield as did limited aeration; however, there is one noteable exception when considering aeration, i.e., the no cultivation, limited aeration, no wheel traffic yield was highest of all. The reasons for this seemingly odd yield prediction are discussed below and illustrate the feedback mechanism in the model and the need to have fairly detailed, physiological based plant growth subroutines in such models. Yields were taken at 170 days after emergence. The model predicted that the no cultivation, no wheel traffic, limited aeration crop

---

[5] R. R. Bruce, Soil physicist, USDA-ARS, Watkinsville, Ga. Personal communication.

Table 2. Predicted cotton yield, high water treatment.[†]

No. cult.	No wheel	Wheel
	————————kg/ha————————	
	No aeration problems	
0	1,645	1,608
3	1,677	1,405
4	1,594	1,433
5	1,633	1,394
	Limited aeration	
0	1,752	1,531
3	1,591	1,351
4	1,519 (1,591)[‡]	1,403 (1,351)[‡]
5	1,591	1,357

[†] Irrigated when the soil matrix potential at 15 cm reached −0.3 bars.
[‡] Numbers in parenthesis were calculated for the fourth cultivation on Day 50 rather than Day 35.

would have more bolls open on Day 170 than any other simulation. The limited aeration had changed the root pattern of this crop, Fig. 8, and root weights, Table 2. Early in the season (20 to 30 days) the root weights, top weights, and leaf areas were essentially the same for the zero cultivation, limited aeration, and fully aerated conditions. As the rooting depth increased, the limited aeration condition became more restrictive and reduced root weights. This resulted in more photosynthate being available for reproductive organs. Thus, more fruiting sites and bolls were produced in the poorly aerated crop than in the well aerated crop. Since water and nutrients were generally not limiting this season, the poorly aerated crop was predicted to have a higher yield.

The numbers in parenthesis in Table 2 illustrate the effects of timing of cultivations, i.e., the fourth cultivation of the bracketed number was on Day 50 after emergence rather than Day 35. With no wheel traffic the effect was predicted to increase yield but with wheel traffic it was predicted to decrease yield. Again these results were due to changes of timing in plant fruiting behavior due to rooting patterns and moisture stress. It can also be noted that the effects of numbers of cultivations are mixed. Under well aerated conditions, three cultivations over the same time period were predicted to give slightly higher yields than five cultivations. Under limited aeration conditions it either didn't matter or the reverse was true. In the model, since infiltration was not changed by tillage or wheel traffic and weeds are not modeled, these responses are due to slight changes in rooting patterns and timing of physiological events.

The simulated root weights at the end of the season for the same treatments of cultivation, wheel traffic, and aeration as in Table 2 are shown in Table 3. In general, the simulations predict that wheel traffic reduces root weight and limited aeration reduces root weight. With good aeration, cultivation is predicted to reduce root weight; however, under limited aeration conditions with no wheel traffic, cultivation may stimulate some increase in root weight. If the fourth of four cultivations is on the 50th day after emergence there are fewer roots than if this cultivation

Table 3. Predicted cotton root weights, high water treatment.[†]

No. cult.	No wheel	Wheel
	g/cm³	
	No aeration problems	
0	36.8	30.0
3	35.3	29.6
4	38.0	31.6
5	35.4	28.7
	Limited aeration	
0	32.8	28.3
3	33.4	28.2
4	34.2 (33.2)[‡]	30.2 (27.9)[‡]
5	33.4	27.6

[†] Irrigated when the soil matrix potential at 15 cm reached −0.3 bars.
[‡] Numbers in parenthesis were calculated for the fourth cultivation on Day 50 rather than Day 35.

Table 4. Predicted cotton yields, low water treatment.[†]

No. cult.	No wheel	Wheel
	kg/ha	
	No aeration problems	
0	1,139	1,066
3	1,200	1,128
4	1,166 (1,178)[†]	1,089 (1,144)[†]
5	1,172	1,128

[†] Irrigated when the soil matrix potential at 45 cm reached −2.3 bars.
[‡] Numbers in parenthesis were calculated for the fourth cultivation on Day 50 rather than Day 35.

were on the 35th day. All of these observations would be expected from field or controlled environmental experiments[3,4] which gives some confidence in the model results.

The yield predictions under the low water treatment are shown in Table 4. The actual lint yield under these conditions was 1,316 kg/ha[5]. It was assumed that aeration would not be a problem under this management system. Again wheel traffic was predicted to decrease yield, but some cultivation was predicted to increase yields. Cultivation on the 50th day after emergence rather than on the 35th day of a four cultivation treatment stimulated yield slightly under both traffic conditions.

The simulated rooting patterns comparing no cultivations to five cultivations on Day 50 after emergence are shown in Fig. 3 for the dry treatment. Under five cultivations there have been some roots pruned in layer 1 column 4, fewer roots in layers 2, 3, and 4 under the plant and not as deep rooting in the 14th layer as the zero cultivation pattern. The root pruning pattern can be seen in Fig. 4 between Days 49 and 50, i.e., the day before and the day of the last cultivation. The roots are gone on Day 50 in layer 1 column 4.

Zero cultivations, no wheel, no aeration problems
Roots in each cell, total

Units — g/cm³ soil

row	1	2	3	4	5	6	7	8	9	10	11
1	2	2	1	0							
2	3	2	2	2	1						
3	7	6	5	3	2	1					
4	4	4	4	3	2	2	0				
5	4	2	2	2	2	2	1	0			
6	2	2	2	2	2	2	1	0			
7	2	2	2	2	2	1	0				
8	2	2	2	2	2	1	0				
9	2	2	2	2	1	0					
10	2	2	2	1	1	0					
11	2	1	1	0							
12	1	1	0								
13	0	0									
14	0										
15											
16											
17											
18											
19											
20											

Total = 7.6419 g dry weight

Five cultivations, no wheel, no aeration problems
Roots in each cell, total

row	1	2	3	4	5	6	7	8	9	10	11
1	2	2	1								
2	3	2	2	2	2	1	0	0			
3	6	6	5	3	2	2	1	0			
4	4	4	3	3	2	2	1	0			
5	2	2	2	2	2	2	1				
6	2	2	2	2	2	1	0				
7	2	2	2	2	2	1	0				
8	2	2	2	2	1	0					
9	2	2	2	1	0						
10	2	1	1	0							
11	2	1	0								
12	1	1									
13	0										
14											
15											
16											
17											
18											
19											
20											

Total = 7.0356 g dry weight

Day 50

Legend

$\checkmark = 0.0000$
$0.0000 < 0 = 0.0001$
$0.0001 < 1 = 0.0005$
$0.0005 < 2 = 0.0050$
$0.0050 < 3 = 0.0100$
$0.0100 < 4 = 0.0150$
$0.0150 < 5 = 0.0200$
$0.0200 < 6 = 0.0250$
$0.0250 < 7 = 0.0300$
$0.0300 < 7 = 0.0350$
$0.0350 < 9 = 0.0400$
$0.0400 < --$

Fig. 3. Root weight profiles for dry treatment comparing cultivation frequencies.

Day 49

Five cultivations, no wheel, no aeration problems
Roots in each cell, total

Units — g/cm³ soil

	1	2	3	4	5	6	7	8	9	10	11
1	2	2	1	0							
2	3	2	2	2	1	1	0				
3	5	5	4	3	2	2	1	0			
4	3	3	3	2	2	2	1	0			
5	2	2	2	2	2	1	0	0			
6	2	2	2	2	1	1	0				
7	2	2	2	2	1	0	0				
8	2	2	2	2	1	1	0				
9	2	2	2	1	1	0					
10	2	1	1	1	1						
11	1	1	0	0	0						
12	1	0	0								
13	0										
14											
15											
16											
17											
18											
19											
20											

Total = 5.5335 g dry weight

Day 50

Five cultivations, no wheel, no aeration problems
Roots in each cell, total

	1	2	3	4	5	6	7	8	9	10	11
1	2	2	1								
2	3	2	2	2	2	1	0	0			
3	6	6	5	3	2	2	1	0			
4	4	4	3	3	2	1	1	0			
5	2	2	2	2	2	2	1	0			
6	2	2	2	2	2	1	0				
7	2	2	2	2	2	1	1				
8	2	2	2	2	2	1	0				
9	2	2	2	1	1	0					
10	2	2	1	1							
11	2	1	1	0	0						
12	1	1	0								
13	0										
14											
15											
16											
17											
18											
19											
20											

Total = 7.0356 g dry weight

Legend

$\not< = 0.0000$
$0.0000 < 0 < \; = 0.0001$
$0.0001 < 1 < \; = 0.0005$
$0.0005 < 2 < \; = 0.0050$
$0.0050 < 3 < \; = 0.0100$
$0.0100 < 4 < \; = 0.0150$
$0.0150 < 5 < \; = 0.0200$
$0.0200 < 6 < \approx 0.0250$
$0.0250 < 7 < \; = 0.0300$
$0.0300 < 8 < \; = 0.0350$
$0.0350 < 9 < \; = 0.0400$
$0.0400 < \; --$

Fig. 4. Root weight profiles for dry treatment before and after cultivation.

Day 50

Units — g/cm³ soil

Five cultivations, no wheel, no aeration problems
Roots in each cell, total

	1	2	3	4	5	6	7	8	9	10	11
1	2	2	1								
2	3	2	2	2	2	1					
3	6	6	5	3	2	2					
4	4	4	3	3	2	2	0				
5	2	2	2	2	2	1	1	0			
6	2	2	2	2	2	1	1	0			
7	2	2	2	2	2	1	1	0			
8	2	2	2	2	1	0	0				
9	2	2	2	1	1	0					
10	2	2	1	1	0						
11	1	1	1	0							
12	0	1	0								
13											
14											
15											
16											
17											
18											
19											
20											

Total = 7.0356 g dry weight

Five cultivations, wheel, no aeration problems
Roots in each cell, total

	1	2	3	4	5	6	7	8	9	10	11
1	2	2	1								
2	3	3	2	2	2						
3	8	8	7	5	4						
4	4	4	4	3	3						
5	2	2	2	2	2		0				
6	2	2	2	2	2	1					
7	2	2	2	2	2	0					
8	2	2	2	2	2	0					
9	2	2	2	2	1						
10	2	2	1	1	0						
11	2	1	1	1	0						
12	1	0	0	0							
13	0										
14											
15											
16											
17											
18											
19											
20											

Total = 9.2321 g dry weight

Legend

≮	= 0.0000
0.0000 < 0	= 0.0001
0.0001 < 1	= 0.0005
0.0005 < 2	= 0.0050
0.0050 < 3	= 0.0100
0.0100 < 4	= 0.0150
0.0150 < 5	= 0.0200
0.0200 < 6	= 0.0250
0.0250 < 7	= 0.0300
0.0300 < 8	= 0.0350
0.0350 < 9	= 0.0400
0.0400 < --	

Fig. 5. Root weight profiles for dry treatment comparing wheel traffic zones.

Five cultivations, no wheel, no aeration problems
Volumetric water content of soil

— Units — cm³/cm³ soil —

	1	2	3	4	5	6	7	8	9	10
1	4	4	2	2	0	0	0	0	0	0
2	3	3	3	4	4	4	4	5	5	5
3	3	3	3	3	3	4	4	4	5	5
4	4	4	4	4	4	5	5	5	5	5
5	5	5	5	5	5	5	6	6	6	6
6	4	5	5	5	6	6	6	6	6	6
7	5	5	5	5	6	6	6	6	6	6
8	5	5	5	6	6	6	6	6	6	6
9	5	5	6	6	6	6	6	6	6	6
10	5	6	6	6	6	6	6	6	6	6
11	6	6	6	6	6	6	6	6	6	6
12	6	6	6	6	6	6	6	6	6	6
13	6	6	6	6	7	7	7	6	6	6
14	7	7	7	7	7	7	7	7	7	7
15	7	7	7	7	7	7	7	7	7	7
16	7	7	7	7	7	7	7	7	7	7
17	7	7	7	7	7	7	7	7	7	7
18	7	7	7	7	7	7	7	7	7	7
19	7	7	7	7	7	7	7	7	7	7
20	7	7	7	7	7	7	7	7	7	7

Total = 315.46 mm water

Five cultivations, wheel, no aeration problems
Volumetric water content of soil

	1	2	3	4	5	6	7	8	9	10
1	4	4	2	2	1	1	1	1	1	1
2	3	3	3	4	4	3	3	3	3	3
3	3	3	3	3	3	3	3	3	3	3
4	4	4	4	4	4	3	3	3	3	3
5	4	5	5	5	5	3	3	3	3	3
6	4	5	5	5	5	3	3	3	3	3
7	5	5	5	5	6	6	6	6	6	6
8	5	5	5	5	6	6	6	6	6	6
9	5	5	5	6	6	6	6	6	6	6
10	5	5	6	6	6	6	6	6	6	6
11	5	6	6	6	6	6	6	6	6	6
12	6	6	6	6	6	6	6	6	6	6
13	6	6	7	6	7	7	7	7	6	6
14	7	7	7	7	7	7	7	7	7	7
15	7	7	7	7	7	7	7	7	7	7
16	7	7	7	7	7	7	7	7	7	7
17	7	7	7	7	7	7	7	7	7	7
18	7	7	7	7	7	7	7	7	7	7
19	7	7	7	7	7	7	7	7	7	7
20	7	7	7	7	7	7	7	7	7	7

Total = 305.19 mm water

Day 50

Legend

∢	= 0.00
0.00 < 0	= 0.05
0.05 < 1	= 0.10
0.10 < 2	= 0.15
0.15 < 3	= 0.20
0.20 < 4	= 0.25
0.25 < 5	= 0.30
0.30 < 6	= 0.35
0.35 < 7	= 0.40
0.40 < 8	= 0.45
0.45 < 9	= 0.50
0.50 <	--

Fig. 6. Water content profiles for dry treatment comparing wheel traffic zones.

	Oxygen concentration in soil No aeration problems										Oxygen concentration in soil Limited aeration									
	Unit — ATM partial pressure																			
	1	2	3	4	5	6	7	8	9	10	1	2	3	4	5	6	7	8	9	10
1	--	--	--	--	9	9	9	9	9	9	7	5	3	1						
2	--	--	9	9	9	9	9	9	9	9	3	0								
3	9	9	9	9	9	9	9	9	9											
4	9	9	9	9	9	9	9	9	9	9										
5	9	9	9	9	9	9	9	9	9	9										
6	9	9	9	9	9	9	9	9	9	9										
7	9	9	9	9	9	9	9	9	9	9										
8	9	9	9	9	9	9	9	9	9	9										
9	9	9	9	9	9	9	9	9	9	9										
10	9	9	9	9	9	9	9	9	9	9										
11	9	9	9	9	9	9	9	9	9	9										
12	9	9	9	9	9	9	9	8	8	8										
13	9	9	9	9	8	8	8	8	8	8										
14	9	9	9	9	8	8	8	8	8	8										
15	9	9	9	8	8	8	8	8	8	8										
16	9	9	8	8	8	8	8	8	8	8										
17	9	8	8	8	8	8	8	8	8	8										
18	9	8	8	8	8	8	8	8	8	8										
19	8	8	8	8	8	8	8	8	8	8										
20	8	8	8	8	8	8	8	8	8	8										

Day 80
Legend
≮ = 0.00
0.00 < 0 < = 0.02
0.02 < 1 < = 0.04
0.04 < 2 < = 0.06
0.06 < 3 < = 0.08
0.08 < 4 < = 0.10
0.10 < 5 < = 0.12
0.12 < 6 < = 0.14
0.14 < 7 < = 0.16
0.16 < 8 < = 0.18
0.18 < 9 < = 0.20
0.20 < --

Fig. 7. Oxygen content profiles for wet treatment comparing well aerated and poorly aerated conditions.

	Roots in each cell Zero cultivation, no wheel, no aeration problems											Roots in each cell Zero cultivation, no wheel, limited aeration											Day 80
	Units — g/cm³ soil																						Legend
	1	2	3	4	5	6	7	8	9	10	11	1	2	3	4	5	6	7	8	9	10	11	≮ = 0.0000
																							0.0000 < 0 < = 0.0001
1	2	2	1	0	0							2	2	1	0	0							
2	3	3	2	2	2	2	1	0	0			3	3	2	2	2	2	1	0	0			0.0001 < 1 < = 0.0005
3	--	--	--	9	6	4	2	2	2	1	0	--	--	--	--	6	4	3	2	1	1	0	
4	--	--	--	9	6	4	3	2	2	1	0	--	--	--	8	6	4	3	2	2	1	0	0.0005 < 2 < = 0.0050
5	4	5	5	5	4	3	3	2	2	1	0	4	4	4	4	3	3	2	2	2	1	0	
6	3	3	3	3	3	3	2	2	2	1	0	3	3	3	3	2	2	2	2	2	1	0	0.0050 < 3 < = 0.0100
7	3	3	3	2	2	2	2	2	1	0	0	2	2	2	2	2	2	2	2	1	0	0	
8	3	3	3	2	2	2	2	2	1	0		2	2	2	2	2	2	2	2	1	0		0.0100 < 4 < = 0.0150
9	3	3	2	2	2	2	2	1	0	0		2	2	2	2	2	2	2	1	0	0		
10	3	2	2	2	2	2	1	1	0			2	2	2	2	2	2	1	0	0			0.0150 < 5 < = 0.0200
11	2	2	2	2	2	2	1	0	0			2	2	2	2	2	1	1	0				
12	2	2	2	2	2	1	0	0				2	2	2	2	1	1	0					0.0200 < 6 < = 0.0250
13	2	2	2	2	1	1	0					2	2	2	1	1	0	0					
14	2	2	2	1	1	0						2	2	1	1	0	0						0.0250 < 7 < = 0.0300
15	2	2	1	1	0	0						2	1	1	0								
16	1	1	1	0								1	0	0									0.0300 < 8 < = 0.0350
17	1	0	0									0	0										
18	0											0											0.0350 < 9 < = 0.0400
19																							
20																							0.0400 < --
	Total = 31.52 g dry weight											Total = 29.99 g dry weight											

Fig. 8. Root weight profiles for wet treatment comparing well aerated and poorly aerated conditions.

Table 5. Cotton yield × equipment size for high water treatment.

No cult.	2 Row	4 Row	6 Row	8 Row
		kg/ha		
		No aeration problems		
0	1,608	1,627	1,633	1,633
3	1,405	1,544	1,589	1,611
4	1,433	1,516	1,539	1,555
5	1,394	1,516	1,555	1,572
		With aeration problems		
0	1,531	1,644	1,683	1,700
3	1,351	1,472	1,511	1,527
4	1,403	1,461	1,480	1,490
5	1,357	1,472	1,511	1,533

Table 6. Predicted cotton yield × equipment size for low water treatment.

No. cult.	2 Row	4 Row	6 Row	8 Row
		kg/ha		
		No aeration problems — Low water		
0	1,066	1,105	1,116	1,122
3	1,128	1,166	1,178	1,183
4	1,089	1,128	1,139	1,150
5	1,128	1,150	1,155	1,161

The simulated rooting patterns comparing no wheel traffic to a 26 cm wide × 31 cm deep traffic zone are shown in Fig. 5. The roots are blocked out of columns 6 to 9 from layers 2 to 6 due to the high bulk density. The total root weights were higher, however, on Day 50 for the wheel traffic area than the non-traffic area. This can be seen in layers 2 and 3, columns 1 to 3. Later the non-traffic roots exceeded the traffic roots as was shown in Table 3. The water content profiles for these same conditions are shown in Fig. 6. There is less water in layers 2 to 6, columns 6 to 10 with wheel traffic than without wheel traffic. This is due to the low value of θ_s for the traffic zone shown in Table 1. Figures 5 and 6 point out that any water or nutrients trapped in a wheel compacted zone will be essentially useless to the crop plant since roots will not penetrate this zone and water movement is reduced within or out of the zone.

The oxygen content of the soil on Day 80 for the wet treatments are shown in Fig. 7. This followed a period of several, frequent irrigations. In the limited aeration case almost all of the oxygen in the profile has been used, while in the good aeration case there is plenty of oxygen for root growth. The resulting root profiles are shown in Fig. 8. There are less roots in the limited aeration case than in the well aerated one. The root distribution is also somewhat different between the two cases. The limited aeration case has fewer roots in layers 4 and 5 and fewer roots deeper in the profile, layers 7 to 17 than the good aeration case. Thus, plants grown under limited aeration conditions will have a different carbohydrate supply balance than well aerated plants, i.e., the root-shoot ratio will be less.

The utility of some of these types of simulations may be seen in Tables 5 and 6. In these tables the cotton yield as a function of equipment size for different water and aeration levels are predicted. As one moves from two row equipment, which usually means a tractor with a tricycle front end, where every row will have a wheel track to eight row equipment where only two rows will have wheel tracks we can see a projected yield increase. If one is interested in a two row, field crop tractor, i.e., one that has the front wheels in the same row as the rear ones, then the four-row column in Table 5 and 6 apply. The upper level of yield approaches the no-wheel traffic case as the equipment size gets larger. One effect, however, not shown in these tables is that as equipment gets larger so does tire size. It is observed in the field[3] that as equipment size gets larger the major effect is tire width and not depth of compaction. Therefore, as this simulation program presently works, tire size would have to increase 5 cm before the results would be detected.

SUMMARY AND CONCLUSIONS

The GOSSYM cotton growth simulation model has been modified to account for hydraulic property changes due to cultivation and wheel traffic, for mechanic impedance due to these same operations and for changes in root growth due to low oxygen stresses. The results have shown that root growth and other physiological processes are altered by these external processes, especially the timing of fruit maturation. The resulting root weights and patterns of development appear reasonable when compared to observation. Such simulations have been combined to estimate the effects of machinery size on predicted yield. Since weed control and water infiltration were not modeled or included in this model, the responses are generally small and are based upon the physiological response of the plant to changes in rooting patterns.

LITERATURE CITED

1. Baker, D. N. 1979. Simulation for research and crop management. p. 533–546. *In* F. T. Corbin (ed.) Proc. World Soybean Conference II, Raleigh, N.C.

2. ————, J. R. Lambert, C. J. Phene, and J. M. McKinion. 1976. GOSSYM: A simulator of cotton crop dynamics. p. 100–133. *In* Computers applied to the management of large-scale agricultural enterprises: Proc. U.S.-U.S.S.R. Seminar, Moscow, Riga, Kishinev.

3. ————, ————, and F. D. Whisler. 1979a. A developmental history of the cotton simulation model GOSSYM. Agronomy Abstr. 71:9. Am. Soc. of Agron., Madison, Wis.

4. ————, J. A. Landivar, and J.R. Lambert. 1979b. Model simulation of fruiting. p. 261–263. *In* Proc. Cotton Prod. Res. Confs. Phoenix, Ariz.

5. ————, ————, F. D. Whisler, and V. R. Reddy. 1979c. Plant responses to environmental conditions and modeling plant development. p. 69–135. *In* W. L. Decker (ed.) Proc. Weather and Agriculture Symp., Kansas City, Mo.

6. Baver, L. D., W. H. Gardner, and W. R. Gardner. 1972. Soil physics. 4th ed., John Wiley and Sons, Inc., New York.

7. Brooks, R. H., and A. T. Corey. 1964. Hydraulic properties of porous media. Hydrology Papers. Colorado State Univ. 3:1–27.

8. Bruce, R. R., and M. J. M. Romkens. 1965. Fruiting and growth characteristics of cotton in relation to soil moisture tension. Agron. J. 57:135–140.

9. Campbell, R. B., D. C. Reicosky, and C. W. Doty. 1974. Physical properties and tillage of paleudults in the Southeastern coastal plains. J. Soil Water Conserv. 29:220–224.

10. Eavis, B. W., H. M. Taylor, and M. G. Huck. 1971. Radicle elongation of pea seedlings as affected by oxygen concentration and gradients between shoot and root. Agron. J. 63: 770–772.

11. Fye, F. E., V. R. Reddy, and D. N. Baker. 1979. The validation of GOSSYM: Arizona conditions. Agronomy Abstr. 71:14.

12. Gardner, W. R., and M. S. Mayhugh. 1958. Solutions and tests of the diffusion equation for the movement of water in soil. Soil Sci. Soc. Am. Proc. 22:197–201.

13. Lambert, J. R., D. N. Baker, and C. J. Phene. 1975. Simulation of soil processes under growing row crops. Paper No. 75-2580. Winter Meeting ASAE, Chicago, IL. 12 p.

14. ————, ————, and ————. 1976. Dynamic simulation of processes in the soil under growing row crops: RHIZOS. p. 258–287. In Computers applied to the management of largescale agricultural enterprises. Proc. U.S.-U.S.S.R. Seminar. Moscow, Riga, Kishinev.

15. Landivar, J. A., D. N. Baker, and J. N. Jenkins. 1979. The application of the cotton simulation model GOSSYM in genetic feasibility studies. Agron. Abstr. 11:13. Am. Soc. of Agron., Madison, Wis.

16. McWhorter, J. C., and B. P. Brooks, Jr. 1965. Climatological and solar radiation relationships. Miss. Agric. Exp. Stn. Bull. 715.

17. Melhuish, R. M., M. G. Huck, and B. Klepper. 1974. Measurement of aeration, respiration rate and oxygen diffusion in soil during the growth of cotton in a rhizotron. Aust. J. Soil Res. 12:37–44.

18. Taylor, H. M., and H. R. Gardner. 1963. Penetration of cotton seedling taproots as influenced by bulk density, moisture content and strength of soil. Soil Sci. 96:153–156.

19. Whisler, F. D. 1976. Calculating the unsaturated hydraulic conductivity and diffusivity. Soil Sci. Soc. Am. J. 40:150–151.

20. ————, J. R. Lambert, C. J. Phene, J. E. Parsons, and D. N. Baker. 1977. Cotton root growth, water uptake and impedance limitations. Agron. Abstr. 69:95. Am. Soc. of Agron., Madison, Wis.

21. ————, J. A. Landivar, and D. N. Baker. 1979a. A sensitivity test of weather variables in the GOSSYM cotton simulation model. Abstr. Workshop on Crop Simulation. p. 35.

22. ————, ————, and ————. 1979b. A sensitivity test of soil, water parameters in the GOSSYM cotton simulation model. Agronomy Abstr. 71:17. Am. Soc. of Agron., Madison, Wis.